Eat To Live Bible

The Ultimate Cheat Sheet &

70 Top Eat To Live Diet Recipes

By: Samantha Michaels

TABLE OF CONTENTS

PUBLISHERS NOTES

Disclaimer

DEDICATION

This book is dedicated to people who desire to have a change of lifestyle through quick and easy dieting.

CHAPTER 1- EATING TO LIVE

Do you know that while food is the culprit for many diseases, food could also be the cure? You do not need to subject yourself to costly treatment and procedures when you can resolve your health problems such as obesity with nutritional intervention.

Diseases Traceable to Poor Nutrition and Excess Weight

Here are just some of the diseases and health conditions that you can trace from poor nutrition and excess weight:

- Acne
- Allergies
- Angina
- Arthritis
- Constipation
- Adult diabetes
- Gallstones
- Gout
- Headaches
- Hemorrhoids
- Hypertension or high blood pressure
- Indigestion
- Kidney stones
- Osteoporosis
- Sexual dysfunction
- Stroke

Nutritional intervention can reverse and prevent the occurrence of these diseases and more. When you eat to live and benefit from it, you can turn your back against costly medicines and procedures. Never underestimate the power of your body for self-healing. However, it needs to receive essential nutrients to activate and sustain this power. The primary and best source of nutrients is food, as long as you make the right choices.

CHAPTER 2- NUTRITION IS POWER

The diet plan believes that you main weapon to lose weight healthily is the wisdom you get from knowing your nutrition. Hence, the book provides you all the information you need so that you will learn to develop the healthy habit of eating to live. The benefit of that habit is sustainable weight loss results along with optimum health.

Here, you will find out why eating high nutrient-density food is favorable to sustainable and healthy weight loss results.

Unrefined Carbs

Carbohydrate is an essential nutrient. In fact, the body depends much on it more than any other nutrients. It is the source of energy. It does not make sense therefore to restrict its consumption, as you will deprive your body of the nutrient that it needs the most.

However, you must also understand that there are bad carbohydrates and good carbohydrates. Bad carbohydrates are rich in sugar and fats as well as empty calories. These are the refined carbohydrates, the ones causing your unnecessary weight, and endangering your health.

Good carbohydrates on the other hand are low in calories, rich in fiber and other nutrients. Unrefined carbs contain lower calories than fats do. You can increase your intake of carb-rich food without necessarily earning calories in excess, unlike fats.

Making your good carbs as the primary source of your calorie requirement is a healthy way to satisfy your hunger. These carbs are also rich in fiber that helps you to feel fuller longer. Therefore, you should not fear good carbs or stay away from it as if it is a plague. Just make sure that you are choosing unrefined carbs.

Examples of Unrefined Carbs

Here are some examples of unrefined carbs:

- Natural whole grains
- Fresh fruits
- Raw Vegetables
- Beans and Legumes

The book will discuss how you can easily distinguish good carbs from bad carbs. Bad carbs are the ones that are discrediting that good reputation of carbohydrates as a nutrient, and if you are to stay away from carbs, it should be bad carbs you are staying away from.

Calories from Carbs and Calories from Fats

It is better to get your calories from unrefined or good carbs than to get it from fats. For instance, eating huge amount of fresh fruits and raw vegetables and receive their calories are much better than consuming dairy products, oils, or meats.

Meats, oils, and dairies are calorie-rich food that it is extremely difficult for your body to resist excessive amount of calories before you can satiate your appetite. If these food types are calorie-dense, fruits and vegetables are nutrient-dense. You can readily satisfy your hunger without consuming excess calories.

Getting your calories from carbs allows your body to burn them efficiently for consumable energy. Calories from fats are more likely to turn into unnecessary body fats that go to the storage and show as bulges or excess weight.

Food Favorable to Weight Loss

Nutrient-dense food is the type that has multiple weight loss benefits. They can fill up your stomach easily without piling calories. These are the following:

- Green vegetables
- Fresh fruits
- Beans and Legumes

Making green vegetables your food base is a surefire way to drop the pounds. According to Dr. Fuhrman, the more greens you eat, the more weight you can lose. He suggests this golden rule in vegetable consumption: eat one pound of raw greens and one pound of cooked (preferably steamed) greens daily. Following this rule will lead you straight to losing your excess weight healthily and sustaining the healthy weight loss results.

Top Food Types According to their Density of Nutrients

Following is the list of food ranked according to the density of nutrients:

1. Raw leafy green vegetables – the darker the color, the greater the density of nutrients
2. Green vegetables – they can be raw, frozen, or steamed
3. Non-green but non-starchy vegetables like mushroom and onion
4. Beans and Legumes
5. Fresh fruits
6. Starchy vegetables like pumpkins and sweet potatoes
7. Whole grains
8. Raw nuts and seeds
9. Fish
10. Fat-free dairy
11. Meats from grass-fed cattle and poultry
12. Eggs

Not included here are food items that measure only a single digit nutrient density or none at all such as red meats (8%) and refined sweets (0%).

Eating vegetables is also beneficial to longevity as it delays the aging process, and prevents degenerative diseases. You can drop the risks by as much as 86% with increase in the consumption of your veggies.

The Importance of Good Fats

Good fats are essential to meet your dietary requirements, and can help your body to lose weight healthily. These essential fatty acids are important for your growth and development and in the treatment and prevention of chronic diseases.

Essential fatty acids are of two main types- Omega 3 fatty acids, and Omega 6 fatty acids. From these two main fatty acids, your body is able to produce other fatty acids or the non-essential fats.

It is important that you have a good balance of Omega 3 and Omega 6 fatty acids.

Too much consumption of Omega 6 fatty acids leads to inflammatory ailments and diseases. Deficiency in Omega 3 fatty acids, meanwhile, can trigger multiple ailments and diseases such as heart, stroke, and depression, autoimmune and skin diseases. Results from clinical studies are also finding the connection between Omega 3 deficiency and cancer.

Further, Omega 3 is anti-inflammatory that can keep Omega 6 fatty acids in check. To improve your health and encourage the loss of your excess weight, it helps that you increase your intake of food rich in Omega 3 fatty acids and reduce your intake of food rich in Omega 6 fatty acids.

Food Sources of Omega 3 Fatty Acids

Here are some of the dietary sources of Omega 3 fatty acids:

- Flaxseed and flax oil
- Soybeans
- Tofu
- Walnuts

Protein Requirements

There is no doubt that protein is an essential nutrient. The only question is how much of it does your body really need to favor healthy weight loss results?

Recommended Amount of Protein

Male   approximately 55 grams daily for a male weighing 150 pound

Female approximately 44 grams daily for a female weighing 120 pound

The World Health Organization recommends that you get 5% of your calories from protein. For people with normal health condition, 2.5% may already be sufficient, according to Dr. Fuhrman. Plant-based food is a good source of protein, contrary to common belief that you should source your protein from animal-based food, as discussed earlier.

CHAPTER 3- BASICS OF HUMAN NUTRITION

Human Nutrition Basics

The foundation of your sustainable weight loss success with this diet plan is to understand human nutrition. You have to realize that your food choices can dictate your weight loss results. In addition, your food choices have the power to lengthen your lifespan or cut it short. Wrong choices result with weight problems such as obesity as well as chronic diseases.

The American Diet

The main culprit in many health issues and weight problems in the country has its roots in the American diet. This is the kind of diet where the food that enters the body has empty calories, which is more popular as junk food.

Americans are fond of eating 'empty-calorie' food, the type that deprives the body of essential nutrients including fiber. What makes matters worst is the fact that most people in the country live a sedentary lifestyle or do not have enough physical activity to burn their caloric intake, empty calories included.

Obesity as the Top Health Problem

It is not surprising that the top health issue confronting Americans today is obesity. Across the world, obesity is also becoming one of the major health problems of the population.

Therefore, there is a continued increase in the population of dieters. These people go on a diet to get rid of their weight problems. The thing is, most of them fail to achieve successful results with their diet. Either this is due to strict restrictions making them to quit early or they regain their initial weight loss and even add more pounds to their original weight prior to dieting.

Losing weight has become a necessity to the population to address obesity. Here are just five of the top complications of this weight disorder:

- Increase in the risk of premature death
- Diabetes
- High blood pressure
- Heart diseases
- Cancer

The Typical Solutions to Obesity

Surgical Procedures

Those with severe obesity resort to surgical procedures to correct their weight problems to avoid the health complications. Surgery, such as the gastric bypass, is a costly treatment and may bring higher risks of adverse effects. The 'Eat to Live' diet plan is the best alternative that can save you from the high cost and risks of surgical procedures.

Diet Programs

Another common solution that the obese and other people with excessive weight take is to diet. Because of this, several diet programs flood the market, each one of them claiming to deliver fast and effective weight loss results to combat obesity and weight problems. However, these diet programs may deliver initial weight loss results, but dieters find that if they go off their programs, they regain their weight and add some more.

Even the most common and basic dieting technique, that is to lower the caloric intake, is ineffective. If you ask why, then here is the simple answer: the volume of your food intake does not matter as much as the quality of food that you take in your body to achieve permanent weight loss. Do you know what makes you gain your excess weight? It is because you eat food deficient in nutrients.

To illustrate, take for instance two persons A and B. Person A gets his calories mostly from junk and processed food. Person B eats the same amount of food as person A, but he chooses to eat nutritious food such as fruits, veggies, lean meat, and fish. With the same volume of caloric intake, who do you think has weight problems to solve?

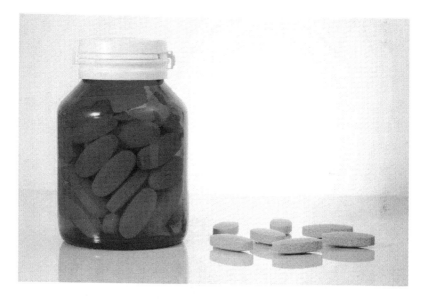

Weight Loss Drugs

You may also notice a flood of weight loss pills and other drugs in the market. While these medicines will allow you to lose weight, they often come with a price- literally and figuratively.

Not only are these medications expensive, you know they can also bring toxic effects to your body. In addition, you must be aware that not because a certain medicine has no side effects, you can claim that it is already completely safe. According to Dr. Fuhrman, from their medical education, doctors learn that all medicines, regardless of their side effects, are toxic albeit in differing degrees.

Metabolism and Genetics

Others will also readily blame their genetics or metabolism as the reason they are overweight. The truth is, notwithstanding the rate of your metabolism is and your genetic make-up, you can always reach your recommended or ideal weight.

How is this possible? Make healthy food choices and increase your physical activity. Several clinical studies prove strongly that the powerful combination of eating healthily and increasing your physical activity is the best way to lose weight regardless if your genes and metabolism.

To lose weight and sustain the results, you will have to give up your fondness of the American diet, which is rich in fats and low in fiber. You will also have to turn away from living a sedentarily and start to become more active physically. These two combined is the only way you can really drop the pounds and keep your ideal weight forever.

The Need to Lose Excess Weight

Losing weight has become a necessity for most people, not because they want to look physically attractive, but because their health condition tells them to do so. Several studies confirm the relation between weight and mortality.

From these studies, results show that those who live longer are people within their recommended weight. Obese people tend to have a shorter lifespan. As these studies prove the existing relation between body weight and mortality, you can say that the heavier you weigh, the greater your chances of premature death.

Your waistline is your gauge. To see quickly if you are at risk, pinch the area near your belly button. For men, you are safe if you are only able to pinch less than half an inch (1/2) of your skin in that area. For women, the safe level is less than one (1) inch. If you can pinch more, then consider yourself a potential candidate for health problems from minor to severe.

To help you determine your ideal or recommended weight, you can use the table below as your easy reference:

Men		Women	
Height: First five feet		**Height:** First five feet	
Weight: 105 pounds		**Weight:** 95 pounds	
Add: Five (5) pounds for every inch subsequently		**Add:** Four (4) pounds for every inch subsequently	
Example:		Example:	
Height	Weight	Height	Weight
5'0	105 lbs	5'0	95 lbs
5'1 (+5)	110 lbs	5'1 (+4)	99 lbs
5'2 (+10)	115 lbs	5'2 (+8)	103 lbs

However, the above is just a guide as each person differs when it comes to body types and bone structure.

Use the body mass index or BMI as an indicator to determine if you are overweight, underweight, or within your normal body weight. Here is the formula:

English	Metric
BMI = weight in lbs / (height in inches)^2 (703)	BMI = weight in kg / (height in m)^2
	Underweight: <18.5 Normal Weight: 18.5 – 24.9 Overweight: 25 – 25.9 Obese: 30 and up

Tip: You can use BMI calculator online.

While both can help you in assessing your weight risks, Dr. Fuhrman suggests using the waist measurement as indicator. Men whose waist measures greater than 40 inches, and women with 36 inches and more have higher risks of health issues including heart attacks.

Pay attention to the fats around your waist more than any other area in your body, as this can indicate your health risk level. If for instance your waist measurement is bigger relative to your body type, i.e. small built or thin, the solution is to work out to build your muscles.

Keep this in mind: It is necessary to lose excess weight, as any additional fat in the body will reduce your lifespan. The only way to get rid of your excess weight through effective and safe diet is to eat the right food that is high in nutrient-density.

Are You Really Hungry?

Most people find it difficult to distinguish real hunger from the opposite. There are those who readily eat thinking they are satisfying their hunger where in fact they are just thirsty and need to drink water. Some will devour their food as source of comfort. Others can hardly control their appetite, and there are those who are dealing with addiction to food.

According to Dr. Fuhrman, the trigger to such disorders and behavior towards food consumption is lack of nutrients in the body. The solution is to improve food choices to resolve nutritional deficiency and promote nutritional excellence. Once you are able to change your diet from junk to high nutrient density, you will be able to overcome all unnecessary hunger, and satisfy only your true hunger.

You can only distinguish real hunger when you start to eat nutritiously. With consumption of high nutrient-density (nutrient per calorie), you can withdraw from your poor habit of craving and eating unhealthy or junk food. After your withdrawal, you will now have the ability to eat less, as you will only have to satisfy your true hunger, enabling you to lose excess weight and sustain it.

Why You Need to Eat Less

Eating less is crucial if you want to prolong your lifespan. Recognizing and satisfying your true hunger is the only effective method to be able to reduce your calorie consumption. Reducing your calorie consumption through satisfying only your true hunger is the only way to enjoy a healthy and longer life.

Consider the following:

*   Several studies including the recent study published by the Proceedings of the National Academy of Sciences confirm that restricting calories with high nutrient food prevent s chronic health conditions and radically increases lifespan.
*   Higher intake of fats from unhealthy food speeds up the aging process and promotes the growth of some tumors in the body. The same studies corroborate the strong association between long life and slimness, and shorter lifespan and obesity.
*   From multiple researches for long life, only one finding remains proven and prevalent. This is the fact that lessening your food consumption provided your body receives essential nutrients will prolong your lifespan. This is the foundation of Dr. Fuhrman's nutrient-density equation for sustainable weight loss: H= N/C.

How to Avoid High Calorie Food and Eat Less

Following the above, the goal is to limit your calorie intake and yet to make sure that your body gets the nutrition it needs for optimum health. Your diet therefore should consist of highly nutritious food items and preferably none of those with poor nutrients.

Eating high nutrient-density food will enable you to get rid of your craving for junk and empty calorie food. This is because highly nutritious food will enable you to meet your recommended nutrition and energy intake while satisfying your true hunger. Food low in calories yet rich in nutrients and fiber fill up your stomach that will prevent you to feel unnecessary hunger to overeat.

How Your Body Systems Work

For better understanding, look at how your body systems work. Your brain is responsible for controlling your drive to eat food. Your digestive system, for its part, is responsible for monitoring your calorie and nutrient intake. It then sends the information it collects to your brain, specifically to the hypothalamus, that controls your hunger drive.

Hence, if your digestive system finds that your body lacks the calories and nutrients it needs, it will keep on sending information to your brain that in turn will signal you to eat more, resulting with overindulging in unnecessary food.

Nutrients and Fiber to Reduce Calorie Intake

Nutrients and fiber will help you to control your appetite to reduce your calorie intake. Dr. Fuhrman recommends that instead of juicing your fruits or vegetables, it is healthier and more beneficial to eat them. Juicing or any form of food processing will only strip the fiber and nutrients from your fruits and veggies. In reducing weight, therefore, it is crucial that you consume your fruits and veggies in their natural form and condition.

There is no shortcut to losing weight through magic pills, powder, or other formula and form. You cannot go on living an unhealthy lifestyle and turn to them as your instant or magical solutions to lose weight. This is true if you want to get rid of your excess weight permanently, safely, with no compromises or trade-offs to your good health condition.

How to Use Your Slow Metabolism to Your Advantage

According to Dr. Fuhrman, you should avoid seeing your slow metabolism as a defect that will prevent you from losing weight successfully. Instead, you should take advantage of it in order to live longer. It only becomes a disadvantage if you continue eating the American Diet- a diet consisting of high calorie and low nutrients.

Besides, slow metabolism is not the cause of your excess weight for the following reasons:

- Your resting metabolism will drop to a certain extent with lower calorie intake, but this drop is hardly significant to prevent weight loss.
- When you do not limit your intake of calories, your resting metabolism reverts to its normal rate. This makes succeeding diets more difficult in achieving weight loss results.
- The cycle of losing and gaining weight for people with weight problems is not due to the speed of the metabolism. The fluctuation is from the starting and stopping the dieting plan or program.

To use your slow metabolism to your advantage, fill your body with food high in nutrients and fiber and low in calories. With such kind of food in your body, your slow metabolism will work in your favor to delay your aging process, contribute to your optimal health, resulting with your longer lifespan.

In fact, people whose genes predispose them to slow metabolic rate are more likely to live longer. However, this is only realizable when they choose high nutrient-density food combined with increased physical activity.

Therefore, the first basic step to lose weight and sustain it is to start to eat to live longer.

CHAPTER 4- ADVANTAGES OF A NUTRITION RICH DIET

Using natural and organic food as a way to improve one's well-being is not a new concept at all.

Ancient people from past civilizations thrived well on a diet of organic and natural grown food. They also learned how to utilize particular plants for medicinal purposes.

Unfortunately, with the emergence of the Industrial period, and the introduction of sugar into our diets, many of us have fallen into the trap of eating "convenient" food items (a.k.a. processed food,) and drinks that have extremely high levels of sugar. Time and time again, these have proven to be detrimental to our health.

Going back to the basics and subscribing to a nutrient rich diet, you can:

1. Lose weight naturally, gradually and safely. Because this diet is high in essential nutrients and low in fat, sugar, and salt, it is relatively easier to shed off the unwanted poundage by simply adhering to the principles of the diet. More importantly, you do no put yourself at risk (or complicate existing health issues, if any) by following a diet that encourages rapid weight loss in a short amount of time.

Although this may sound counterintuitive, gradually losing weight over a longer period is actually a far better option than dropping excess pounds in just a few weeks or days. For starters, crash dieting would only compromise the health of your internal organs: your liver and kidneys, in particular. This could lead to lifelong and potentially dangerous health complications like: all manner of liver diseases and kidney failure.

Secondly, crash dieting is often interpreted by the body as a period of starvation: a state wherein your internal organs experience extreme nutrient deprivation. Because there is no longer an external source of nutrients, the body turns inward. It burns off its store of reserved fat in the adipose tissues. In extreme cases, the body can even burn off muscle cells to acquire a protein source. This is the real reason why people lose weight when they crash diet.

Unfortunately, once the diet is over, the body autonomically (without conscious thought on your part) takes extreme measures to ensure that it does not undergo starvation again. The first thing it does is to create voluminous amount of adipose tissues around the abdominal area, the thighs and the butt. Any fat or oil you eat afterwards are deposited into these tissues, and are reserved for later use.

This makes it easy to rapidly regain all the weight you've lost before, and then some! Worse, because the body is "reserving" the stored fat, it becomes harder to lose the excess pounds afterwards.

With a nutrient rich diet though, your weight loss is safer and definitely more sustainable in the long run.

2. Helps stabilize your blood sugar. This is especially beneficial to diabetics or those who are already in the pre-diabetic stage. A nutrient rich diet discourages the excessive consumption of table sugar or other sweeteners that can alter the balance of insulin in the bloodstream.

An imbalance of insulin can lead to other health issues like: hyperglycemia (high levels of blood sugar due to low insulin in the bloodstream,) hypoglycemia (low levels of blood sugar due to excessive insulin intake or production,) blindness, obesity, and kidney failure, etc. It can also complicate existing heart, liver, and kidney conditions.

For people with Type I diabetes, uncontrolled hyperglycemia increases the risk of diabetic coma; while uncontrolled hypoglycemia can lead to insulin shock that could lead to diabetic coma as well. This is a life-threatening condition that would need medical intervention.

3. Helps stabilize your blood's cholesterol level. A diet high in oils and fats can dramatically increase your cholesterol level. This can cause: liver and kidney failure, obesity, diabetes, etc. Unfortunately, having high cholesterol also causes 2 life-threatening conditions, namely: heart failure (congestive cardiac failure) and stroke (cerebrovascular accident.)

With a nutrient rich diet though, the intake of oils and fats are limited. This lowers all the risks normally associated with high blood pressure due to unhealthy blood cholesterol level.

At the same, this diet does not completely removed oils and fats from its prescribed meals, as these are also nutrients essential to keep hair, nails and skin healthy. The key is to find healthier options or alternatives, and to keep portions small.

Rediscovering the taste and benefits of naturally grown and organic food can improve your health greatly. These recipes are incredibly easy to make too. Aside from helping you save a lot of money, these recipes are proven to yield tasty and delicious food and beverages each and every time.

So if you are ready, here are 70 recipes you might want to incorporate in your daily meals.

CHAPTER 5- FIGHTING OBESITY BY EATING THE RIGHT FOOD

In losing weight, you do not simply target to drop your excess pounds. You also have to consider your health condition aiming to live longer. You now know that Dr. Fuhrman's formula expressed in the equation H = N/C is crucial to a longer life.

You also realize that the key in using this formula for longer life is to increase your intake of food with high nutrient density and decrease your intake of unhealthy food, especially those with empty nutrients and low in fiber.

To appreciate and apply this formula more effectively, it will help if you understand why you need to follow it in the first place. To do this, here are some of the common practices in human nutrition along with the effects of these practices to weight and health.

Advantages and Disadvantages of Having a Sweet Tooth

It is human nature to have a sweet tooth. It is for this reason that this human nature can also be an advantage when losing weight healthily. Your sweet tooth can direct you to one of the types that has high nutrient-density- fruits.

Several studies prove that in consuming fruits, you are equipping your body with the strongest shield against some types of the deadly 'C', namely: lung, pancreatic, prostate, esophageal and oral cancers. Fruits are indispensable in benefiting from a healthy weight loss diet.

Other health benefits of fruits are the following:

* Delays the aging process
* Slows brain deterioration

- The strongest weapon in lowering mortality rate from cancers

The disadvantage of sweet tooth is that most people choose to satisfy it with commercial products filled with refined sugar. Refined sugar is a major contributor to excess weight and all its complications. Clearly, to benefit from having a sweet tooth, redirect your craving to eating fresh fruits.

The Problem with Refined Food

The problem with sugar and other refined food items is that while they encourage your sweet tooth and increase your food consumption, they are deficient in nutrients and fiber. Unregulated consumption of the food results with weight gain, and deficiency in nutrients and fiber triggers multiple health problems.

With scientific and medical proof of their link, some of these health problems are the following:

- Cancers specifically oral, colorectal, stomach, breast, intestinal, thyroid, and respiratory tract
- Diabetes
- Heart diseases
- Gallbladder

Do you have to give up eating refined food or dishes with refined food ingredients? While it is preferable, Dr. Fuhrman suggests that you do not have to give it up completely. However, it is better to choose these food items carefully, and limit their intake.

For instance, if you have to eat pasta, choose whole-grain or bean pasta and then use only small quantity as much as possible. In its place, you can increase your green veggies such as lettuce, kale, spinach as well as onions, mushroom, and tomatoes. As much as you can, avoid those that have white flour as the main ingredient.

Food items that undergo refining are substantially deficient in nutrients. While you can see food products carrying the labels "enriched" or "fortified", they are still deficient in nutrients and the enrichment and fortification are only about ten percent (10%) that usually comes from adding synthetic vitamins and minerals.

Clearly, your food choice matters in losing weight. Choose food items that are as close to their natural condition as possible to enjoy healthy weight loss.

Not All Whole Food is Healthy

Nevertheless, you should also be aware that not all whole-grain food is healthy. There are those that involve processing already stripping them of their nutrients and health benefits. For instance, whole grain cereals have already lost most of their nutritional value from processing, as well as finely ground whole wheat.

Following this, if you are to buy your oatmeal, avoid the instant variety as they no longer contain the nutritional value associated with oats. To avoid cooking, you can soak your oats for a certain number of hours before consumption.

Relationship between Fats from Unhealthy Carbs and Your Waistline

Do you know that the surefire way to become obese is to eat food that is a mixture of fats, sugar, and starch or flour? Fats stimulate the appetite. The more you eat fatty food, the more you will crave for it. The thing is the body is quick to convert these fats from food into body fat. Unnecessary fats in the body become your excess weight and your health risk.

Unnecessary fats mixed with blood sugar are dangerous. Not only will this combination result with excessive weight gain, it will also interfere with the normal functions of your body systems. This interruption, in turn, has a negative impact on your otherwise normal health condition. This is the reason obesity has become the top health problem in the country. Most people are addicted to food rich in fats and refined carbohydrates.

Consume Raw Fruits and Vegetables

To curb your addiction to fatty food and those loaded with refined carbs, turn your sweet tooth to consuming raw fruits and vegetables, as Dr. Fuhrman suggests. They can give your body the highest level of protection against diseases, specifically cancers. In fact, you can say that raw fruits and vegetables are the best anti-cancer diet that can help you lose your excess weight fast and safely.

Here is the recommendation of Dr. Fuhrman to get maximum health benefits from your diet: it should consist of huge amount of raw food, next in line are cooked food items or ingredients with lesser caloric density, and then last in the hierarchy are calorie-rich cooked starchy veggies and grains. Choose healthy cooking method as well, such as boiling or broiling, and avoid using oil as it can inhibit your weight loss.

Fiber's Role

Fiber is an important element in the health equation, as it is vital to your nutrition. Food enriched with fiber allows you to lose weight and protects your body against cancer and heart diseases. It is also crucial that you source your fiber from fruits and vegetables, whole grains, beans, and raw nuts and seeds to experience the best health benefits. Again, it is a matter of food choice.

From numerous studies, you will find that fiber is a necessary component in losing weight and preventing diseases. Food items that cause diseases have a common denominator: they are deficient in fiber. Therefore, you may conclude that if you increase your intake of fiber-rich food, especially raw and natural food, you can expect to drop the pounds and live a longer life altogether.

Chapter 6- Natural Weight Loss Pills

In Dr. Fuhrman's 'Eat to Live' book, you will discover a natural pill that works like magic in delivering both weight loss and health benefits. No, it is not one of those popular weight loss pills in the market, the product of media hype.

Phytochemicals Deliver Magical Results

This magic pill is phytochemicals. They are not literally pills in the sense that you source them from your vegetables or plant food. Contrary to common belief that you cannot source your protein from plants, the fact is 25% of the calories in your vegetables belong to protein.

Just like fruits, veggies are also excellent sources of nutrients specifically phytochemicals. They function best when you use them as natural base in satisfying your nutritional need. Vegetables contain carbohydrates, fats, and protein, and unlike other carb and fat sources, you get high nutrient density from veggies. Additionally, the carb and fats you get from veggies are the healthy types- such as complex carbs and essential fatty acids.

Phytochemicals work as a tough shield to protect your body against diseases, including cancer. In fact, studies show that these compounds inhibit the formation of cancer cells in the body. Since vegetables and fruits are powerhouse sources for phytochemicals and other nutrients, eating more of these food types can speed up your healthy and sustainable weight loss.

Dr. Fuhrman also suggests sourcing your phytochemicals from fresh fruits and vegetables rather than taking them as dietary pills. You will benefit more from natural food than from pills. This is because most of the compounds and nutrients you will find working in synergy in natural food, you can hardly extract and isolate without compromising their nutritional value and health benefits.

Authorities Recommend Consumption of Natural Fruits and Vegetables

According to the book, eight top health organizations who are authorities and experts in the field unite in encouraging the population to eat fruits and vegetables in their natural condition. These authorities also promote consumption of plant-based food rather than animal-based food.

These health organizations are the following:

- American Heart Association (AHA) Nutrition Committee
- American Cancer Society
- American Academy of Pediatrics
- American Diatetic Association
- American Society for Clinical Nutrition
- National Institutes of Health, Nutrition Research Division
- Council on Epidemiology and Prevention
- Council on Epidemiology and Prevention

One of their findings is that animal-based food is one of the top causes of various life threatening diseases including obesity.

Another reason to benefit from plant-based food is that the phytochemicals can flush out toxins from the body and prevent cell damage. They can also serve as both the body's soldier and armor against cancerous cells. In short, nutrients from fresh fruits and vegetables are the best protection you can give your body so that you can enjoy your normal weight and excellent health condition.

Here are just ten of the numerous natural chemicals in fruits and vegetables that promote optimum health and work as anti-cancer:

- Anthocyanin
- Cathechins
- Flavonoids
- Isoflavones
- Pectins
- Phenolic Acids
- Polyphenols
- Protease Inhibitors
- Saponins
- Sterols

Difference between Plant-Based Food and Animal-Based Food

In losing weight and preventing diseases, it is not only because you are eating fats from animal-based food, or eating empty calorie. It is also about depriving your body of phytochemicals and other nutrients from plant-based food.

Briefly, plant-based food has higher nutrient-density than animal-based food. Dr. Fuhrman illustrates this by comparing the nutrient density of broccoli (plant-based food) and steak (animal-based food). Which do you think between the two has more protein?

If you were to base your answer from common belief, you would choose steak. However, do you know that broccoli has almost twice as much protein as steak? Consider this: broccoli has 11.2 grams of protein, and steak has only 5.4 grams.

You also know that calories should come from carbohydrates, protein, and fats. The difference between the two is that animal-based food earns its calories from fats, while calories from plant-based food are mostly protein.

Lesson from Popeye

Popeye gets his strength from spinach, and he was right. Indeed, the greens can deliver power punches. Notice how all the big animals eat greens as their main food source.

What this means in the context of healthy weight loss is that by increasing your intake of plant-based food and decreasing animal-based food, you can satisfy the protein requirement of your body to maintain and promote optimum health while losing excess weight.

Yes, Popeye made the right choice in spinach, as green vegetables are the hands-down winner when it comes to nutrient-density. If this were a scoring competition, it would hit the perfect score, 100%.

Learn from Popeye, if you wish to lose weight and be strong like him, your main dish should be green salad. You can mix and match green leafy vegetables and devour it with peace in your mind knowing that you are doing your body a favor- weight wise and health wise.

Why Weighing Food and Eating Smaller Portions Are Ineffective

It is an exercise in futility to weigh your food and eat smaller servings. This is because the digestive system does not mind the weight of the food when sending signals to the brain receptors. It is the bulk or volume, the calories, and the nutrients present in the food that matter

Therefore, you can eat the same weight or servings of junk food and high nutrient-density food and get your weight loss results from the latter. You have come to know that your green leafy vegetables have the highest nutrient-density among all food types.

Animal Protein May Have Negative Effect On Weight Loss

The book suggests going easy on animal protein, as it may have negative effects on the weight loss process. It lists several studies that confirm the following:

• The connection between animal protein and formation of cancer cells in the body

• How animal protein can encourage heart diseases which normally people will attribute to fats and cholesterol

From the above, it is easy to conclude to skip or avoid eating animal-based food. However, this is not what the book means.

Strictly Plant-Based Diet Does Not Guarantee Healthy Weight Loss

Avoiding animal-based food and shifting to a strict plant-based food diet is not a guarantee for healthy weight loss results. There is a big difference between eating an exclusively vegetarian diet and a diet rich in fresh fruits and vegetables.

You can become a vegetarian and still fail to get healthy weight loss results. On the other hand, you can still eat animal-based food; consume fruits and vegetables in large amount, and enjoy healthy weight loss benefits.

It all boils down to food choices. While statistics corroborate that vegetarians have longer lifespan, it is important to note that these vegetarians choose natural plant-based food over refined and processed plant-based food.

Demystifying the Animal Protein Myth

Perhaps until now, you still believe that you can only get your protein from animal-based food, and that you need this protein for faster growth and development. This is a myth that you should learn to debunk.

If you continue to believe this myth, you will still equate healthy nutrition with having a larger built and experiencing early maturity, and that faster growth means excellent health. According to Dr. Fuhrman, the reverse is true and several studies support this.

Faster growth and early maturity speed up the aging process. They are more of risk factors for deadly diseases like cancer than they are indicators of growth. Therefore, the slower one's growth and maturity is, ruling out nutritional disorders and serious diseases, the more likely it is to live longer.

Revisiting Various Diet Plans To Lose Weight

Since the top health problem plaguing the American society today is obesity, it is understandable that countless of weight loss products and diet plans flood the market. After all, weight loss has become a big business, and who does not want to earn profitable money?

Obesity is a serious health problem. Ignoring it can lead to fatal complications. However, most diet plans are products of hype capitalizing on the need of the population to lose weight.

Some of these plans are successful, but you can measure success in a number of ways, such as the following: popularity, how much weight as well as how fast you can lose it with the program, and how well the program works according to its claim.

Yet, you have to consider one crucial and perhaps the most important factor in measuring success: does the diet plan allow you to lose weight healthily? By healthily, it means that:

-        The diet plan is able to deliver permanent weight loss

-        Results are safe with no adverse effects on your body or health condition

-        The diet promotes your overall good health

You cannot say that a diet plan is successful regardless of how much and how fast you can lose your weight with it if the results are temporary and especially if it compromises your health.

The Danger with Popular and Trendy Diet Plans

Sadly, diet plans that enjoy huge popularity and sales are also the riskiest. These plans promote quick weight loss solutions, something that is irresistible to the majority of people who wish to drop their pounds. These plans are giving the people what they want to hear, and not what they need to hear.

It is therefore important that you do your part in revisiting these diet plans and weigh them according to their advantages and disadvantages, pros and cons, merits and demerits. Use the health equation and all the information you find from this guide as your easy reference.

If you read the entire book by Dr. Fuhrman, you will see all the evidence and proof why the health equation or Dr. Fuhrman's formula for weight loss is the only way to achieve healthy weight loss results. Again, the formula is H = N/C, and that there is no shortcuts or substitute to eating a diet of high nutrient-density to lose weight and enjoy healthy and sustainable weight loss results.

The bottom line in evaluating diet plans and programs is if they are not able to deliver permanent or sustainable weight loss without compromising your health, then it is best to avoid the program. In resolving your weight loss problem, you can never and must not separate health and weight.

Why You Should Avoid These Weight Loss Plans

It is best that you save your hard-earned money instead of getting into these popular diet plans. These plans cater to the American Diet and they will never go against it. Hence, you get diet plans that are rich in fats, and deficient in fiber.

The majority of the pre-packaged food that usually accompanies the diet plan lacks essential nutrients. Therefore, you may lose weight but you will also put your health at risk.

If you do your part, you will also learn that these programs will deliver initial weight loss results, but succeeding weight loss will depend on your continued patronage of the program. In short, you are married to the program. If you decide to cut ties or divorce from it, you regain your weight and you even add more. This is dangerous in the long run.

How Different the Eat to Live Diet Plan Is

Healthy weight loss is simple: you have to learn to eat to live. Unlike other diet plans, this diet by Dr. Fuhrman encourages you to eat as much as you can, as long as you are making the right food choices. You have learned from earlier discussion that right food choice is the one that has high nutrient-density or rich nutrient-per-calorie. You have also learned that natural plant-based food is the highest in nutrient-density.

This is how the 'Eat to Live' diet is different from the rest. It may not cater to the love of the society for food rich in fats but low in fiber, but it is the only method that can deliver sustainable weight loss results without health risks or trade-offs. It is a diet plan that focuses on your health first, and then permanent weight loss results will come naturally. You eat to live, and by doing so you normalize your weight and keep it long lastingly.

It is also easier to follow the diet plan since there are no strict restrictions unlike most plans. You do not even have to count calories, weigh your food, or control the serving size. Your body systems will do that for you. If you fill your body with high nutrient-density food, you digestive system will automatically communicate with your brain to signal you to stop eating as you have already satisfied your true hunger.

CHAPTER 7- TIPS FOR EATING GREENS

Fresh, crisp, green salads should constitute the majority of your meals. These are incredibly healthy, tasty, and have very few calories. This is why you can consume as much of these as you want. You can even use these as side dishes for meals, or slide these in with your favorite sandwiches and burgers for a heftier snack.

In order to keep things healthy though, here are a few ground rules:

1. Fresh vegetables are always the best. Although you can use frozen or canned food items like carrots, corn, green peas, etc. as substitutes for fresh produce, you have to remember that some of the essential nutrients may have been lost through the processing stage. Limit the use of frozen or canned food items as much as possible.

    If you have to use these, thaw frozen vegetables completely in the fridge to preserve its flavor. If using canned vegetables, wash these thoroughly under cool, running water to remove excess starch, salt and sugar.

2. Raw and fresh fruits are better than canned and bottled ones. Raw and fresh fruits provide 100% more nutrients than their processed counterparts. These also contain less sugar.

    If you would like to incorporate fruit juice into your salads, always use freshly squeezed ones. And squeeze these just before you are about to serve or eat the salad. Use fruit pulp and zest whenever possible.

3. Be adventurous when it comes to trying out new fruits and vegetables. If you limit yourself to the few produce you

know, chances are: you will end up with a boring diet. There are numerous fruits and vegetables you can try that will help you eat healthier and even offer you new dishes to love. So be brave when it comes to eating healthier.

4. Keep your salad dressings simple and separate. Your healthy salads could quickly turn into an unhealthy bowl of processed food if you keep using ready-made condiments, dips and dressings. To adhere to the principles of this diet, you need to learn how to make your own dips and dressings from scratch. This is the only way to ascertain exactly what goes into your own food. You can also find healthier alternatives to sugar, and control the amount of salt or oil when you make your own meals.

CHAPTER 8- MAKE YOUR OWN SALAD

Broccoli and Cauliflower Salad

Ingredients:

2 heads	Broccoli, washed, drained, cut into bite-sized florets
1 head	Cauliflower, washed, drained, cut into bite-sized florets
-	Garlic Power
-	Olive Oil
-	Salt to taste
-	White Pepper to taste (Optional)
-	Water

You would also need: Ice Bath (a bowl of water and ice cubes)
Strainer or Colander
Slotted Spoon

Directions:

1. Halfway fill a deep saucepan with water. Allow water to boil over medium heat.
2. Gently slide in the broccoli and cauliflower florets into the boiling liquid. Turn the heat off and put the lid on. Allow the vegetables to sit in the cooking water for only a minute.
3. With a slotted spoon, remove vegetables immediately out of the water and into the ice bath. This stops the cooking process and would help keep the broccoli and cauliflower crisp and vibrant looking. Drain vegetables into a strainer after 2 minutes.
4. Transfer the broccoli and cauliflower florets into another bowl and season with the remaining ingredients. Serve immediately.

Cold White Bean Salad

Ingredients:

For the stew:

2 cups	Dried Small White Beans, washed and drained well (You can substitute dried cannellini or navy beans.)
6 cups	Water
1 cup	Black or Green Olives in brine, washed thoroughly under running water, drained well, pitted then halved
-	Salt to taste
-	Black Pepper to taste

For the emulsion:

¼ cup	White Wine or Apple Cider Vinegar
½ cup	Extra Virgin Olive Oil
2 tbsp.	Dijon or Yellow Mustard
2 pieces	Garlic Cloves, peeled and grated

For the salad:

1 head	Lettuce of your choice, leaves washed individually, drained well, leaves torn into bite size pieces
2 pieces	Tomatoes, medium-sized, washed, and cut into wedges

You would also need: slow cooker

strainer or colander

wire whisk

Directions:

1. Set the slow cooker at its lowest setting. Place the beans and water in. Cook for 8 hours, or until beans are soft enough to be mashed between your fingers. Carefully pour the beans into a strainer. Wash beans under running water. Drain well before transferring to a large glass bowl.

2. In a separate bowl, combine all the ingredients of the emulsion. Whisk vigorously until liquid turns cloudy. Pour this over the beans and toss gently. Add the green onions and olives. Season to taste with salt and pepper. Chill for at least an hour before serving.

3. To assemble: line a plate with torn lettuce leaves and a few pieces of tomato wedges. Ladle a generous serving of the chilled white beans on top. Serve with bread or soup.

Green Beans and Asparagus Salad

Ingredients:

½ pound	Fresh Snap Beans, washed, ends and strings removed. (You can substitute Edamame, Romano or French Green Beans.)
½ pound	White Asparagus, choose the ones with thicker steams, washed, tough ends snapped off and sliced roughly the same length of the snap beans
5 pieces	Garlic cloves, peeled and minced
1 tbsp.	Olive Oil
-	Water
-	Salt to taste
-	Pepper to taste

You would also need: Strainer or Colander
Slotted Spoon

Directions:

1. In a skillet, heat the oil until slightly smoky. Quickly sauté the minced garlic until it turns lightly brown and aromatic. Remove skillet from flame and allow to cool at room temperature. Keep a careful eye on the garlic as this could still burn. If the garlic seems to be getting browner, simply remove the garlic pieces with a slotted spoon and set aside.

2. In a deep stock pot, add the string beans and cover with water. Bring to a gentle boil, about 3 minutes. Quickly toss in the asparagus slivers and allow to cook for about 1 minute more. Quickly but carefully drain the vegetables into the strainer.

3. Toss drained vegetables into the garlic oil and coat the vegetables well. If you have separated the minced garlic

from the oil, simply toss the toasted garlic on top of the coated vegetables. Season with salt and pepper. Serve immediately.

Mushroom Salad in Teriyaki Sauce

Ingredients:

1 tbsp	Extra Virgin Olive Oil
½ pound	Canned Button Mushrooms, washed under running water, drained well, sliced into thin silvers
¼ cup	Teriyaki Sauce

For the salad:

3 handfuls	Salad greens of your choice, washed, drained and torn into bite size pieces, chilled
1 cup	Grape tomatoes, washed, stems removed, halved, chilled
1 piece	Cucumber, washed, ends removed, cubed, chilled
1 stalk	Celery, washed, roots trimmed, coarsely chopped, chilled

For the garnish:

1 stalk	Scallion, washed, roots trimmed, coarsely chopped
1 tsp	Sesame Seeds, toasted on a dry frying pan, cooled to room temperature

Directions:

1. Place the oil on the skillet and turn the heat up on its highest setting. Caramelize the mushroom slices for about 2 minutes. Turn the heat down.
2. Carefully pour the teriyaki sauce into the hot skillet and allow this to simmer for a few seconds. Remove from flame and allow to cool completely. Chill the sauce for an hour, if possible.
3. Toss all the salad ingredients well into a large glass bowl. Toss with the mushrooms and the teriyaki sauce. Garnish with scallions and sesame seeds just before serving.

Warm Broccoli and Mushroom Salad

Ingredients:

1 head	Fresh Broccoli, large-sized, washed, dried and cut into bite-sized florets
1 tsp	Extra Virgin Olive Oil
¼ cup	Onion, peeled and finely diced
2 tbsp	Balsamic Vinegar
3 cups	Canned Button Mushrooms, washed under running water, drained well, quartered
-	Water
-	Salt

You would also need: Slotted Spoon
                         Ice Bath (bowl of water and ice)

Directions:

1. Halfway fill a small saucepan with water. Add a pinch of salt and set it over high flame. Wait for the water to boil before sliding in the broccoli florets. Cover saucepan. After

2 minutes, remove the vegetables from the hot water with a slotted spoon. Place broccoli immediately into the ice bath to stop the cooking process. Set aside.

2. In a skillet, heat the oil until slightly smoky. Stir fry the onions until wilted, about 1 minute. Carefully pour in the balsamic vinegar. Let this cook until the liquid is reduced in half. This should take no more than 30 seconds. Remove skillet from flame.

To assemble: in a serving or casserole dish, toss the mushrooms and the drained broccoli florets together. Season with more salt, if needed. Serve immediately.

CHAPTER 9- CONDIMENTS, DIPS & DRESSINGS

Avocado and Tomato Dip

Ingredients:

4 cups	Fresh, Ripe Cherry Tomatoes, washed, halved
1 piece	Very Ripe Avocado, medium-sized, stone removed, flesh scooped out
2 tbsp	Freshly squeezed Lemon or Lime Juice, include as much of the pulp as possible, seeds removed
1 handful	Fresh Cilantro or Parsley, washed and dried
-	Salt to taste
-	Pepper to taste

You would also need: Food Processor or Blender

Directions:

Process all the ingredients together, except the salt and pepper. Pulse 5 to 7 times for a chunky consistency. Process for 20 seconds or more for a smoother dip. Chill before serving.

Avocado and Grapefruit Chunky Dip

Ingredients:

1 piece	Overripe Avocado, medium-sized, stone removed, flesh scooped out, then mashed roughly with a fork
1 piece	Grapefruit, peeled, membranes and seeds removed, shred the pulp using your fingers, reserve any grapefruit juice
1 piece	Japanese Cucumber, medium-sized, washed, dried, ends and seeds removed, minced
2 pieces	Celery Stalks, leaves removed, roots trimmed, minced
-	Salt to taste
-	Pepper to taste

Directions:

Combine everything in a bowl and season well. Chill before serving. This is perfect for toasted pita bread or vegetable sticks.

Babaganouj

Ingredients:

2 pieces	Large Eggplants, washed and dried, stems trimmed
½ cup	Tahini
¼ tsp	Cumin Powder
1 piece	Garlic Clove, peeled
1 dash	Garlic Powder
1 dash	Onion Powder
-	Olive Oil or any Vegetable Oil

You will also need: Oven
Baking Dish
Food Processor or Blender
Spoon
Fork

Directions:

1. Preheat oven to 400°F or 200°C.
2. Lightly brush baking dish with olive oil. Place eggplants whole on the baking surface and bake for 30 to 45 minutes. The flesh of the eggplant should be completely soft. Test by pricking with a fork.
3. Remove eggplants from oven and allow to cool completely at room temperature. Once cooled, cut the eggplants lengthwise. Using a spoon, carefully scrape off the seed row and discard. Scrape the remaining flesh into the food processor, along with the remaining ingredients.
4. Process to desired consistency. Some people like this chunky, while others prefer a smoother dip. If you like the chunkier version, process only for a few seconds. A longer processing time will create a thicker but smoother paste. Serve warm.

Berry and Tahini Dip

Ingredients:

4 cups	Cherry Tomatoes, washed and halved
1 cup	Fresh or Frozen Blueberries (You can substitute fresh or frozen strawberries, or any berries that are in season.)
3 tbsp	Tahini or Almond Butter
1 tbsp	Freshly Squeezed Lemon or Lime Juice
1 handful	Fresh Cilantro or Parsley, coarsely chopped
-	Salt to Taste
-	Pepper to Taste

You would also need: Food Processor or Blender

Directions:

Except for the salt and pepper, process everything until you have a smooth consistency. Chill for at least 15 minutes before using. Season to taste. Add more lemon juice if desired.

Blueberry Dip

Ingredients:

1 ½ cups	Fresh or Frozen Blueberries, washed and drained well
2 pieces	Fresh Dates, pitted, washed and drained well
1 tbsp	Apple Cider Vinegar
1 tbsp	Freshly Squeezed Lemon, use as much of the pulp as possible
-	Salt to Taste
-	Pepper to Taste

You would also need: Food Processor or Blender

Directions:

Process all the ingredients, except for the salt and pepper. Transfer to a glass container and chill for at least 15 minutes. Season well with salt and pepper just before serving.

Dill and Tomato Dip

Ingredients:

2 pieces	Ripe Tomatoes, washed and halved
1 tsp	Tahini or Almond Butter
2 tsp	Freshly Squeezed Lemon Juice
1 dash	Fresh or Dried Dill
-	Salt to Taste
-	Pepper to Taste

You would also need: Food Processor or Blender

Directions:

Process everything except the salt and pepper. Add more dill or tahini, desired. Season well with salt and pepper. Chill before serving.

Garlic and Ginger Salad Dressing

Ingredients:

1 knob	Fresh Ginger Root, thumb-sized, washed, peeled and grated, stringy parts discarded

2 pieces	Garlic cloves, peeled and grated
4 tbsp	Light Soy Sauce
4 tbsp	Water
4 tsp	Freshly Squeezed Lemon Juice
2 tsp	Apple Cider Vinegar

You would also need: Grater

Directions:

In a bowl, mix all the ingredients together. Pour over salad greens only before serving.

Herb Dip with Sun Dried Tomatoes

Ingredients:

1 cup	Fresh Mint Leaves, lightly packed, washed, stems removed
½ cup	Fresh Cilantro Leaves, lightly packed, washed
¼ cup	Fresh Chives, lightly packed, washed, roots removed, choose only the green part of the leaves
4 pieces	Sun Dried Tomatoes, coarsely chopped
3 tbsp	Water
2 tbsp	Freshly squeezed Lemon Juice
-	Salt to Taste
-	Pepper to Taste

You would also need: Food Processor or Blender

Directions:

Place the mint leaves, cilantro, chives and tomatoes into the food processor. Add the lemon juice and blend until you have a smooth consistency. If the dip is too thick, add water 1 tablespoon at a time. Season well with salt and pepper. Serve immediately.

Hot Pepper Dressing

Ingredients:

1 piece	Fresh Jalapeño Pepper or bird's eye pepper, washed, stalk removed (If you prefer a milder taste, carefully remove the seeds of the pepper. If not, leave the pepper intact.)
1 piece	Garlic Clove, peeled, halved
1 piece	Onion, small-sized, peeled, sliced into wedges
1 knob	Fresh Ginger, thumb-sized, washed, peeled
1 cup	Water
-	Salt to Taste
-	Pepper to Taste

You will also need: Food Processor or Blender

Directions:

Place everything in the food processor and blend well. You can use this as salad dressing or sauce for meals. Adjust seasoning according to taste.

Mango Dip

Ingredients:

4 pieces	Tomatoes, large-sized, washed, dried, coarsely chopped
2 pieces	Ripe Mangoes, medium-sized, stones removed,

	flesh scooped out
4 tbsp	Apple Cider Vinegar
½ cup	Water
-	Salt to Taste
-	White Pepper to taste

You would also need: Food Processor or Blender

Directions:

Except for the salt and white pepper, simply process all the ingredients together until you have a smooth consistency. Season according to taste and chill before serving.

CHAPTER 10- HOMEMADE BREAD RECIPES

When creating breads from scratch, it is essential to learn at least one basic recipe. This will make it easier to create more complicated breads in the future.

Basic Artisan Bread Recipe (BABR)

With this recipe, you do not need to:

- ❖ Proof yeast,
- ❖ Knead the dough,
- ❖ Punch down the dough and let it rise again,
- ❖ Worry that dough did not rise high or long enough,
- ❖ Worry that dough will fall flat after a long rising period, and
- ❖ Make a new batch of dough every time you need bread.

Ingredients:

5 ½ cups   Whole Wheat Flour

2 ¼ cups	All-Purpose Flour (Plus a few more for dusting)
1 ½ tsp	Granulated or Fast Acting Yeast
1 tbsp	Rock or Kosher Salt
4 cups	Lukewarm Water
2 cups	Cornmeal for dusting
1 dash	Whole seed mixtures for topping, like: Raw Caraway Seeds, Flaxseed, Raw Sunflower Seeds, Poppy and Sesame Seeds

You would also need: Large Mixing Bowls
Saran Wrap
Oven
Baking Sheets or Silicone Mat
Parchment Paper
Pizza Peel or Upside Down Baking Sheet
Kitchen Shears
Baking Stone or Thick-Bottomed Baking Sheet
Metal Broiler Tray
Pastry Brush
Sharp Knife
Wire Rack
Wire Whisk
Wooden Spoon (Optional)

Directions:

1. In a large bowl, whisk together the 2 flours, yeast and salt.
2. Make sure that the water is still slightly warmer than body temperature before adding this to the dry ingredients.
3. Mix flour and water using a wooden spoon, or preferably with your clean, bare hands. Make sure that everything is evenly incorporated and that there are no longer any patches of dry flour.
4. Cover the bowl loosely with saran wrap. Let the dough rise. Set this in a warm, dry place where it can remain

undisturbed for at least 2 hours. However, the longer this dough remains undisturbed, the better its flavor becomes.

5. After the preferred rising period, transfer the loosely covered bowl into the fridge. The dough will retain its "freshness" for the next 14 days. This step also makes the dough easier to work with. Whatever you do, do not punch down the dough, or you will end up with an incredibly dense bread.

6. If you are ready to bake, prepare your pizza peel by dusting it generously with cornmeal. If you are using a baking sheet, turn this upside down and line the bottom with silicone mat or with parchment paper. Dust the surface with cornmeal.

7. Retrieve the chilled dough from the fridge. Remove the cover and lightly dust the surface of the dough with all-purpose flour.

8. Pull up a grapefruit size (approximately 1 pound) dough, and cut it off using your kitchen shears. Add a little more flour to both your hands and the dough so that you can shape it into a ball.

9. Gently tug on the surface of the dough and tuck it underneath. This will help create the crust of the bread. Repeat this step as often as needed to form a smooth surface. This process should be done quickly. About 20 to 40 seconds is enough. Otherwise, all the air inside the dough would escape, and the bread will become dense and chewy after baking.

10. Now shape the dough into a loaf, and place it bottom side down on the pizza peel. Let this rest for another 90 minutes. Cover dough loaf loosely with saran wrap. If the saran wrap seems to be sticking to the surface of the dough, dust again with a light layer of flour.

At this stage, the dough will look like it is deflating. Do not panic. The bread will start to rise once it gets into the hot oven.

11. 30 minutes before baking, heat the oven up to 400°F or 200°C. Place the baking stone on the middle rack of the oven. If you don't have a baking stone, just use a thick-bottomed baking sheet large enough to accommodate the size of your dough loaf.

12. Place an empty metal broiler tray on the rack underneath the baking stone.

13. Just before baking, diagonally score the surface of the dough with a sharp knife. Two ¼ inch deep slashes would suffice.

14. Using a pastry brush dipped in a small amount of water, gently brush off the excess flour on the surface of the dough.

15. Lightly sprinkle your preferred whole seed mixture topping.

16. Prepare 1 cup of hot water before you open the oven door.

17. Very carefully but quickly transfer the dough from the pizza peel and unto the hot baking stone. If you have to gently move the dough with your hand, do so but be careful of the hot surfaces. If you are using parchment paper or silicone mat, simply tug the paper or mat unto the baking stone.

18. Pour the cup of hot water into the empty metal broiler tray underneath the baking stone.

19. Quickly close the oven door to trap in the steam and continue baking the bread at this temperature for 30 to 40 minutes. The bread is done when the crust turns golden brown.

20. Remove the bread from the oven. Allow this to cool completely on a wire rack before slicing or serving it. This will help create a crisp crust but soft interior. Cooling time may take between 15 to 45 minutes, depending on room temperature.

Basil and Pine Nut Loaf

This recipe follows the Basic Artisan Bread Recipe (BABR), but includes these...

Ingredients:

2 tsp	Fresh Basil Leaves, washed, dried, stems removed, and leaves coarsely chopped (You can substitute 1 tsp. of dried basil. Do not use dried basil powder
4 tsp	Raw Pine Nuts, coarsely crushed

From the BABR recipe, you need to remove this:

Dash	Whole seed mixtures for topping, like: Raw Caraway Seeds, Flaxseed, Raw Sunflower Seeds, Poppy and Sesame Seeds

Directions:

1.   Follow BABR directions. But in step 2, incorporate basil leaves and crushed pine nuts with the lukewarm water.
2.   Proceed with the steps 3 to 14.
3.   Skip step 15, then proceed with the remaining steps.

Black and Green Olive Loaf

This recipe follows the Basic Artisan Bread Recipe (BABR), but includes these...

Ingredients:

4 pieces	Black Olives in brine or oil, washed thoroughly, dried, pits removed and minced

| 4 pieces | Green Olives in brine or oil, washed thoroughly, dried, pits removed and minced |

Do not use fresh olives as these become bitter during the long rising process.

From the BABR recipe, you need to remove this:

| Dash | Whole seed mixtures for topping, like: Raw Caraway Seeds, Flaxseed, Raw Sunflower Seeds, Poppy and Sesame Seeds |

Directions:

1. Follow BABR directions. But in step 2, incorporate the olives with the lukewarm water.
2. Proceed with the steps 3 to 14.
3. Skip step 15, then proceed with the remaining steps.

Easy Onion Sourdough Loaf

This recipe follows the Basic Artisan Bread Recipe (BABR) but includes these.

Ingredients:

| 4 tsp | Fresh Leeks, choose only the green parts, washed and minced |
| 1 tbsp | White Wine Vinegar |

From the BABR recipe, you need to remove this:

| Dash | Whole seed mixtures for topping, like: Raw Caraway Seeds, Flaxseed, Raw Sunflower Seeds, Poppy and Sesame Seeds |

1.   Directions:
2.   Follow BABR directions. But in step 2, incorporate the leeks and the white wine vinegar with the lukewarm water.
3.   Proceed with the steps 3 to 14.
4.   Skip step 15, then proceed with the remaining steps.

Garlic and Parsley Loaf

This recipe follows the Basic Artisan Bread Recipe (BABR), but includes these.

Ingredients:

2 tsp	Dried Garlic Granules (Do not use garlic powder or garlic salt.)
2 tsp	Fresh Parsley Leaves, washed, tougher stems removed and minced (You can substitute dried parsley leaves. Do not use dried parsley powder)

From the BABR recipe, you need to remove this:

Dash	Whole seed mixtures for topping, like: Raw Caraway Seeds, Flaxseed, Raw Sunflower Seeds, Poppy and Sesame Seeds

Directions:

1.   Follow BABR directions. But in step 2, incorporate the garlic granules and fresh parsley with the lukewarm water.
2.   Proceed with the steps 3 to 14.
3.   Skip step 15, then proceed with the remaining steps.

Garlicky Loaf

This recipe follows the Basic Artisan Bread Recipe (BABR), but includes these...

Ingredients:

| 1 tsp | Fresh Garlic Cloves, peeled, grated |
| 1 tsp | Dried Garlic Granules (Do not use garlic powder or garlic salt.) |

From the BABR recipe, you need to remove this:

| Dash | Whole seed mixtures for topping, like: Raw Caraway Seeds, Flaxseed, Raw Sunflower Seeds, Poppy and Sesame Seeds |

Directions:

1. Follow BABR directions. But in step 2, incorporate the fresh garlic and the garlic granules with the lukewarm water.
2. Proceed with the steps 3 to 14.
3. Skip step 15, then proceed with the remaining steps.

Raisin and Cashew Loaf

This recipe follows the Basic Artisan Bread Recipe (BABR), but includes these.

Ingredients:

| 3 tbsp | Raisins (You can substitute prunes, but make sure that the pits are removed. Mince the prunes.) |
| 2 tsp | Raw Cashew Nuts, pounded or coarsely crushed |

From the BABR recipe, you need to remove this:

Dash	Whole seed mixtures for topping, like: Raw Caraway Seeds, Flaxseed, Raw Sunflower Seeds, Poppy and Sesame Seeds

Directions:

1. Follow BABR directions. But in step 2, incorporate the raisins and crushed cashew nuts with the lukewarm water.
2. Proceed with the steps 3 to 14.
3. Skip step 15, then proceed with the remaining steps.

Rosemary and Thyme Loaf

This recipe follows the Basic Artisan Bread Recipe (BABR), but includes these.

Ingredients:

2 tsp	Fresh Rosemary Leaves, washed, dried, stems removed, and needles roughly minced (You can substitute 1 tsp. of dried rosemary. Do not use dried rosemary powder.)
2 tsp	Fresh Thyme Leaves, washed, dried, stems removed, and leaves roughly minced (You can substitute 1 tsp. of dried thyme. Do not use dried thyme powder.)

From the BABR recipe, you need to remove this:

Dash	Whole seed mixtures for topping, like: Raw Caraway Seeds, Flaxseed, Raw Sunflower Seeds, Poppy and Sesame Seeds

Directions:

1. Follow BABR directions. But in step 2, incorporate the herbs with the lukewarm water.
2. Proceed with the steps 3 to 14.
3. Skip step 15, then proceed with the remaining steps.

Stove Top Flat Bread

You can very easily create your own flat bread without the need to bake these in the oven. Here is the basic recipe for stove top flat bread.

Follow the Basic Artisan Bread Recipe (BABR), but remove these:

2 cups	Cornmeal for dusting (plus a few more for later)
Dash	Whole seed mixtures for topping, like: Raw Caraway Seeds, Flaxseed, Raw Sunflower Seeds, Poppy and Sesame Seeds

You would also need:	Rolling Pin
	Thick Bottomed Skillet
	Spatula
	Knife
	2 Tea Towels or Thick-Weaved Paper Towels
	Air Tight Container

Directions:

1. Follow BABR steps # 1 to # 5.
2. Skip step # 6.
3. Follow steps # 7 to # 8.

4. Once you have formed a ball, generously sprinkle a flat working surface with flour. Roll your dough into a thick log, then divide this into 8 equal sized disks.

5. With your hand cupped over the dough disks, roll these in a circular motion on the flat working surface until you have a relatively round ball of dough in your hand. Set aside.

6. Repeat step # 5 until you have rolled out 8 dough balls.

7. Dust flour on the rolling pin. Take one dough ball and roll it to a quarter inch thick disk. Make sure that you do not roll the dough too thinly. Add more flour if the dough sticks to the working surface or the rolling pin.

8. Repeat step # 7 until you have rolled out all the 8 dough balls.

9. Heat the skillet on the stove top at the highest setting. Wait for it to become slightly smoky.

10. Meanwhile, take an airtight container and line it with a tea towel. Reserve the other tea towel as "cover." If you are using paper towels, line the airtight container with one sheet. Every time a flat bread is cooked, line it with another paper towel.

11. When the skillet is smoky, place one dough disk at the center of the heated surface. Cooking time should take no more than 1 minute on both sides. Flip the bread using a spatula. If and when the bread starts developing air pockets, simply press down on the dough with your spatula. The bread is done when you see patches of brown spots at the center of the flat bread. The edges should also be firm and not spongy.

12. Remove the cooked flat bread from the skillet and place immediately on the towel lined container. Cover with the other tea towel. If you are using paper towels, you need another sheet to cover the cooked bread. This step actually helps cook the breads further without subjecting it to direct heat. Without this step, the flat bread becomes brittle and chewy.

13.   Repeat steps # 11 and # 12 until all the dough disks are cooked.

Cover the last bread with the tea towel or paper towel and seal the container. Serve immediately or while the flat breads are still warm.

CHAPTER 11- LEGUME AND GRAIN-BASED MEALS

Bean and Corn Salad

Ingredients:

1 cup	Boiled Kidney Beans*
1 cup	Boiled Pinto Beans*
1 cup	Boiled Corn Kernels*
1 piece	Yellow Bell Pepper, washed, cored, seeds removed, diced
1 piece	Red Bell Pepper, washed, cored, seeds removed, diced
2 pieces	Garlic Cloves, peeled, crushed (For a milder taste) or minced (for a sharper taste)
1 handful	Fresh Cilantro Leaves, washed, coarsely chopped
1 piece	Lemon, medium-sized, halved, seeded and juiced
1 tsp	Cumin Powder (You can substitute All Spice Powder)
¼ cup	Soy Oil
-	Salt To Taste
-	Pepper to Taste
1 piece	Jalapeño Pepper, diced (Optional)

*You can substitute canned beans but wash these first to remove excess salt and sugar

Directions:

Toss all the ingredients together in a large bowl. Make sure that you coat everything well with the soy oil. Adjust the taste by adding more salt and pepper, if needed. Use the jalapeño pepper if you want a spicier salad. Chill for an hour before serving.

Beans and Peas Salad

Ingredients:

1 pound	Frozen Green Peas, completely thawed in the fridge, drained
1 pound	Chickpeas, cooked, drained (You can substitute Canned Chickpeas.)
1 pound	Kidney Beans, cooked, drained (You can substitute Canned Kidney Beans)
5 pieces	Garlic Cloves, peeled and minced

For the vinaigrette:

3 tbsp	Extra Virgin Olive Oil
3 tbsp	Apple Cider Vinegar
-	Salt to Taste
-	Pepper to Taste

Directions:

In a large skillet, heat the oil up until slightly smoky. Add the minced garlic until it is lightly brown and aromatic. Remove quickly from the fire. Toss in the green peas, chickpeas, and kidney beans. Mix well. Season to taste, and serve immediately.

Carrot with Cashew in Rice Stew

Ingredients:

2 tbsp	Extra Virgin Olive Oil
3 pieces	Onions, medium-sized, peeled and minced
½ head	Cabbage, medium-sized, washed, julienned, drained well
1 piece	Carrot, medium-sized, peeled and minced
1 piece	Apple, large-sized, washed, peeled, cored and minced
½ cup	Raisins
½ cup	Whole Raw Cashew Nutes
8 cups	Vegetable Stock
¼ cup	Tomato Paste
½ cup	Brown, Red or Wild Rice, washed under running water until the liquid turns clear, drained well
-	Salt and Pepper to Taste
-	Water as needed

Directions:

1   In a Dutch oven, heat the oil over medium flame. Sauté the onion until it turns transparent. Add in the cabbage and carrots. Stir fry for a minute or until shredded cabbage turns limp.

2   Carefully pour the vegetable stock and the tomato paste into the Dutch oven. Turn the flame up to its highest setting and bring the stew to a boil.

3   Add in the washed rice, cashew nuts and the apple slices. Reduce heat to the lowest setting and stir to prevent the rice from sticking to the bottom of the pot. Place the lid on and allow this to simmer until the rice is cooked through. This should take between 20 to 35 minutes.

4   After 20 minutes, check the rice if it can easily be mashed by a fork. Continue with the next step if the rice is soft enough. If not, continue cooking for another 5 minutes. There should still be a lot of liquid in the pot. However, if the rice has absorbed most of the liquid, add water one cup at a time to prevent the stew from drying out.

5   Stir in the raisins. Season with salt and pepper. Remove stew from flame. Serve warm.

Couscous and Vegetable Salad

Ingredients:

1 cup	Couscous, choose the easy-to-cook or instant variety
1 cup	Boiling Water
1 head	Cauliflower, washed, dried and cut into florets
1 piece	Carrot, medium-sized, washed, peeled, diced
1 piece	Green Bell Pepper, small-sized, washed, seeded, diced
1 piece	Shallot, peeled, minced
1/3 cup	Canned Chickpeas, wash under running water, drained well
½ cup	Sun Dried Tomatoes

¼ cup	Black Olives in brine, coarsely chopped
½ tsp	Allspice powder
½ tsp	Cumin Powder
-	Zest of 2 Lemons
2 pieces	Lemons, halved and squeezed, seeds removed
¼ cup	Extra Virgin Olive Oil
-	Salt to Taste

You would also need: Saran Wrap

Ice Bath (bowl of water and ice)

Strainer or Colander

Slotted Spoon

Directions:

1. In a large heat-resistant bowl, combine boiling water and couscous together. Stir once, then cover the bowl with saran wrap. Set aside for 5 minutes to let the couscous absorb the liquid. After 5 minutes, remove the cover and fluff the couscous by gently working it with a fork. Cover with saran wrap again and set aside.

2. In a small pot, place the sun dried tomatoes and cover with an inch of water. Allow the water to simmer gently for 5 minutes. Fish the tomatoes out of the pot and chop these into large chunks. Set aside to cool.

3. Meanwhile, halfway fill a separate pot with water. Allow water to boil before carefully sliding in the diced carrots and cauliflower florets. Cook for 5 minutes. With a slotted spoon, remove the vegetables from the pot and place it into the ice bath.

4. Place the couscous in a tight mesh strainer. Wash under running water to remove excess starch. Drain the couscous well before using.

5.   In a large bowl, toss all the vegetables and spices together, including the lemon juice, lemon zest, and extra virgin olive oil. Season well with salt. Add the couscous and mix gently but thoroughly. Refrigerate until you are ready to serve.

De-constructed Cheese-less, Nut-Free Pesto Pasta

Ingredients:

1 pound	Spinach and/or Carrot flavored Fettuccine, cooked according to package instructions, drained well. Drizzle oil to prevent the strands from sticking together. Set aside. (You can substitute any regular ribbon pasta, such as: Linguine, Pappardelle, Tagliatelle, or Spaghetti.)
1 handful	Fresh basil leaves, washed, stems removed, coarsely chopped
2 pieces	Fresh tomatoes, large-sized, halved, seeds removed, julienned
5 pieces	Garlic cloves, peeled and grated
2 tbsp	Balsamic Vinegar
¼ cup	Extra Virgin Olive Oil
-	Salt to Taste
-	Pepper to Taste

You would also need: Whisk

Directions:

1.   Combine oil with balsamic vinegar and minced garlic. Whisk well until it emulsifies a little. Add the chopped basil leaves. Season with salt and pepper.

2.   Pour this emulsion into the cooked pasta and toss gently but thoroughly. Make sure that each strand of the pasta is well coated with the flavored oil.

3.   Just before serving, add the tomato slices and toss again. Serve immediately.

Fusilli in Garden Salad with Apple Cider Vinaigrette

Ingredients:

2 cups	Spinach and/or Carrot Fusilli or Corkscrew Pasta, cooked according to package directions, drizzled with a little olive oil to prevent pasta strands from sticking together. Chill in the fridge as soon as possible.*
1 head	Broccoli, washed and cut into florets
1 piece	Large carrot, washed, peeled, sliced roughly into the size of the cooked pasta, chilled
1 piece	Large Cucumber, washed, ends removed, skin left intact, sliced roughly the size of the cooked pasta, chilled
1 piece	Large Tomato, washed, sliced in half to remove seeds, sliced roughly the size of the cooked pasta, chilled
-	Water

* You can substitute plain fusilli, or other cut pasta like: Cavatappi, Gemelli, Penne, or Ziti.

For the vinaigrette:

2 tbsp	Olive Oil
2 tbsp	Apple Cider Vinegar

1 tsp	Dried Basil
-	Salt to Taste
-	White Pepper to Taste
You would also need:	Ice Bath (a bowl of water and ice cubes)
	Wire Whisk
	Slotted Spoon

Directions:

1   Halfway fill a small saucepan with water. Bring to boil. Gently slide in the broccoli florets and carrot slices. Cook covered for 1 minute. Turn heat off. With a slotted spoon, scoop vegetables into the ice bath to stop the cooking process. Set aside.
2   In a bowl, whisk together all the ingredients for the vinaigrette until mixture emulsifies slightly. Season to taste.
3   Just before serving, remove the broccoli and carrots from the ice bath and drain well. In another bowl, combine the cooked pasta and all the vegetables. Drizzle the vinaigrette over the pasta and toss. Serve immediately.

Harira (Lentil and Bean Soup)

Ingredients:

1 tbsp.	Extra Virgin Olive Oil
2 pieces	Onions, medium-sized, peeled and minced
2 tsp.	Curry powder
1 tsp.	Ground Cumin Powder
2 tsp.	Fresh Rosemary Leaves, washed, needles coarsely chopped (You can substitute 1 tsp. dried rosemary leaves or dried rosemary

	powder.)
1 tsp.	Fennel Seeds
6 cups	Vegetable Stock
1 cup	Dried White Beans
1 cup	Dried Red Lentils
½ cup	Brown, Red or Wild Rice
1 piece	Tomato, medium-sized, sliced in half to remove seeds, diced
½ cup	Fresh Cilantro Leaves, reserve a few sprigs for garnish later, washed, stems removed, coarsely chopped
1 tbsp.	Tomato Paste
-	Salt to Taste
-	Black Pepper to Taste
-	Tabasco Sauce to Taste
-	Water as needed

Directions:

1   Heat oil in a Dutch oven. Cook onions until transparent. This would take no more than 5 minutes. Add the herbs and spices: cumin, curry powder, fennel seeds and chopped rosemary leaves.

2   Pour in the vegetable stock, along with the beans, lentils and rice. Turn the heat on its highest setting and bring the stew to a boil, uncovered. Stir frequently to ensure that the rice and legumes do not stick to the bottom of the Dutch oven.

3   Once the stew comes to comes to a boil, turn the the heat down to the lowest setting, and allow stew to simmer gently for the next 30 minutes, partly covered. After 30 minutes, test rice, beans and lentils if these are cooked through and can be mashed by a fork. If not, add more water, one cup at a time to keep the stew from drying out.

4   If the rice and legumes are cooked, add in the diced tomatoes and tomato paste. These only have to be heated through. Season well with salt, pepper and Tabasco sauce. Remove from flame. Garnish with cilantro sprigs just before serving.

Lo Mein, Mushroom and Vegetable Salad

Ingredients:

1 pound	Dried Lo Mein Noodles, cooked according to package directions then set aside to cool completely at room temperature.*
3 tbsp.	Roasted Sesame Oil
1 head	Broccoli, medium-sized, washed, dried and cut into florets
4 pieces	Bok Choy Leaves, washed, roots trimmed, sliced diagonally into ½ inch slivers
1 piece	Carrot, large-sized, washed, peeled and

	julienned
1 piece	Red Bell Pepper, medium-sized, washed, cored, seeded and julienned
4 large	Fresh Portobello Mushrooms, stalks removed, caps sliced thinly**
¼ cup	Water
2 tbsp.	Light Soy Sauce
2 tbsp.	Apple Cider Vinegar
3 tbsp.	Sesame Seeds, freshly toasted on a frying pan just before using.

*You can substitute soba noodles or any dried, round egg noodles. Spaghetti also works fine in a pinch.

**You can substitute fresh button mushrooms. Canned portobello and button mushrooms can be used in a pinch. Wash these thoroughly under running water and drain well before using.

You would also need: Strainer or Colander

Directions:

1   Drizzle sesame oil on the cooked noodles. With clean hands, gently massage the oil into the cooked noodles to loosen strands up. Set aside.
2   Fill a saucepan halfway with water. Bring to boil. Turn heat down, then add all the vegetables all at once: broccoli, bok choy leaves, carrots and the sliced portobello mushrooms. Boil for only 2 minutes.
3   Drain the vegetables using the strainer. Discard the cooking liquid. Gently let running water wash over the cooked vegetables to ensure that these remain crisp when you eat them. Drain well.

4 In a large glass bowl, combine the the noodles, cooked vegetables and the remaining ingredients and toss lightly. Chill before serving.

Tabouli

Ingredients:

½ cup	Bulgar Wheat (Choose the easy-to-cook or instant variety)
1 cup	Water
1 piece	Shallot, small-sized, peeled, washed and diced
1 piece	Green Bell Pepper, washed, stem removed, cored, seeded and diced
1 piece	Tomato, large-sized, halved to remove seeds, diced
1 handful	Flat Leaf Parsley, washed, stems removed, chopped roughly
1 piece	Lemon, juiced, seeds removed
2 tbsp.	Extra Virgin Olive Oil
-	Salt to Taste

Directions:

1 In a saucepan, combine the bulgar wheat and the water. Mix well then bring to a boil. Reduce the heat after 3 minutes. Remove the saucepan from the fire then cover with a lid. Let this stand undisturbed for the next 20 to 30 minutes. Remove the lid afterwards to speed up the cooling process. Once the saucepan is cool enough to touch, transfer this to the fridge to cool further.

2   Meanwhile, combine the rest of the ingredients in a large glass bowl. Mix well and season with salt. Chill this mixture in the fridge until the bulgar is ready.

3   To serve, toss the chilled bulgar with the mixed vegetables. Serve as a side dish or as a meal in itself.

Wari Muth (Black Beans with Garan Masala)

Ingredients:

2 cups	Dried Black beans, washed, dried
6 cups	Water
½ tsp.	Dried Red Pepper Flakes
2 tsp.	Fennel Seeds, roughly pounded into a coarse paste
1 tsp.	Garam masala (You can substitute curry powder.)
1 tsp.	Ground Ginger Powder
2 tsp.	Turmeric Powder
-	Salt to Taste
¼ cup	Fresh Cilantro Leaves, coarsely chopped for garnish

You would also need: Slow Cooker

Directions:

Except for the salt and the garnish, place everything in the slow cooker and stir once. Place a lid on and set the cooker on the highest setting. After an hour, turn the cooker down to its lowest setting. Allow beans to cook undisturbed for the next 7 hours. Season well with salt just before serving. Top with chopped cilantro leaves.

CHAPTER 12- SOUP & STEW RECIPES

Asparagus, Corn, and Mushroom Soup

Ingredients:

1 pound	Asparagus, washed, tough stalks trimmed, cut into
1 ½ inch	Long Silvers
2 tbsp.	Extra Virgin Olive Oil
1 ½ cups	Fresh corn kernels, washed and drained*
1 piece	Red Bell Pepper, medium-sized, washed, stem and seeds removed, minced
5 pieces	Fresh Portobello or Button Mushrooms, medium-sized, stalks removed, caps sliced into thin slivers
4 pieces	Leeks, choose the white parts only, washed, roots removed, minced
2 cups	Vegetable Stock
½ tsp.	Curry Powder
-	Salt to Taste
-	Ground Pepper to Taste
-	Water, if needed

*You can substitute canned corn kernels but wash these thoroughly under running water to remove excess starch, salt and sugar. Drain well.

Directions:

1. In a saucepan, heat oil until slightly smoky. Lower flame to medium setting. Stir-fry corn and bell pepper until pepper is limp, or about 3 minutes.
2. Add in the leeks and mushrooms. Stir fry for another minute.
3. Add the curry powder, vegetable stock and the asparagus. Season with salt and pepper. Add more water (half cup at a time) if you find the stock too heavy, too dense or salty.

Allow the soup to boil for 10 minutes at the lowest flame setting, covered.

4. Remove saucepan from flame. Let this stand for 2 minutes to allow the corn to cook through.

5. Ladle soup into bowls and serve immediately.

Barley and Mushroom Stew

Ingredients:

1 cup	Pearl Barley, washed and drained well*
8 cups	Water
2 tbsp.	Extra Virgin Olive Oil
1 piece	Onion, large-sized, peeled and sliced thinly
½ oz.	Dried Porcini Mushrooms, washed then soaked in water for at least 12 hours. Discard soaking liquid. Squeeze dry, then soak for another 30 minutes in water just before using.**
2 tbsp.	Tomato Sauce
1 pieces	Carrot, large-sized, washed, peeled, diced
16 stalks	Celery, large-sized, washed, roots trimmed, finely chopped
-	Salt to Taste
-	Black Pepper to Taste
2 tbsp.	Fresh Parsley, washed, coarsely chopped, for garnish

*You can substitute scotch barley, brown or wild rice, or buckwheat groats.

**You can substitute dried shitake mushrooms.

You would also need:	Slow Cooker Tight Mesh Strainer or Cheese Cloth

Directions:

1. Set the slow cooker on the lowest setting. Pour in the barley and the water. Place the lid on.
2. In a skillet, heat the oil and stir fry the onion slices until these are limp. This should take no more than 3 minutes. Set aside.
3. Meanwhile, squeeze the mushrooms dry. Discard the stems and roughly chop the caps before setting aside. Take its soaking liquid and run it through the tight mesh container (or cheesecloth) to remove the impurities. Set the liquid aside but discard the impurities.
4. Carefully incorporate the cooked onions (and its oil,) the chopped mushrooms, the soaking liquid, carrots, celery and tomato sauce into the pearl barley. Stir gently and place the lid back on. Cook the stew 8 hours.
5. Season to taste before serving. Garnish with chopped parsley.

Black Bean Vegan Soup

Ingredients:

For the salsa:

2 tbsp.	Fresh Cilantro, washed, coarsely chopped
2 tbsp.	Fresh Lemon or Lime Juice, use as much as the pulp as possible
1 piece	Jalapeño Pepper, washed, stem and seeds removed, chopped
1 piece	Ripe Mango, washed, peeled, stone removed, flesh diced
1 piece	Shallot, medium-sized, peeled, washed and minced
-	Salt to taste

-     Pepper to taste

For the soup:

2 tbsp.	Extra Virgin Olive Oil
2 pieces	Onion, large-sized, peeled and minced
1 piece	Carrot, medium-sized, washed, peeled and minced
1 piece	Celery stalk, washed, roots trimmed, minced
4 pieces	Garlic cloves, peeled and minced
½ tsp.	Ground Coriander Powder
½ tsp.	Ground Cumin Powder
3 cups	Vegetable Stock
½ cup	Freshly squeezed Orange Juice, use as much of the pulp as possible
1 can (15oz)	Black Beans, rinsed thoroughly under running water, drained well
-	Salt to taste

For the garnish:

¼ tsp.	Ground Black Pepper Powder, as garnish
1/8 tsp.	Dried Red Pepper Flakes, as garnish
You would also need:	Food Processor or Blender

Directions:

1. To make the salsa: combine all the ingredients in a large bowl. Adjust seasoning according to taste. Chill.
2. To make the soup: heat oil in a large thick bottomed saucepan. Add the garlic, onion, celery and minced carrots all at once. Sauté until celery is limp and diced onions are transparent, or about 3 minutes.
3. Except for the salt, add all the remaining soup ingredients into the saucepan and allow soup to boil uncovered at medium flame. After 5 minutes, remove saucepan from heat. Adjust the taste by seasoning it with salt.

4.   Transfer the contents of the saucepan into the food processor and blend well.
5.   To serve: ladle a small measure of soup into a bowl. Top with salsa and serve with bread.

Corny Nutty Chili Stew

Ingredients:

2 tbsp.	Extra Virgin Olive Oil
2 pieces	Onion, medium sized, peeled, coarsely chopped
2 pieces	Garlic Cloves, peeled, coarsely chopped
2 pieces	Celery Stalks, washed, roots trimmed, coarsely chopped
1 piece	Green Bell Pepper, washed, stem and seeds removed, coarsely chopped
1 cup	Fresh or Frozen Corn Kernels, washed and drained well
1 can (28oz.)	Tomato Juice
1 can (15oz.)	Tomato Sauce
1 cup	Water
1 can (15oz.)	Kidney Beans, washed under running water, drained well
1 cup	Raw Cashew Nuts, whole, washed and drained well
1 cup	Dark Raisins
½ tsp.	Chili Powder (Add more for a spicier stew)
½ tsp.	Tabasco Sauce (Add more for a spicier stew)
1 tsp.	Ground Cumin Powder
1 piece	Bay Lead, whole
1 tbsp.	Fresh Oregano, washed, stems removed, coarsely chopped*
1 tbsp.	Fresh Basil, washed, stems removed, coarsely chopped**
-	Salt to Taste

-          Black Pepper to Taste

*You can substitute 1 tsp. dried oregano leaves. Do not use dried oregano powder.

**You can substitute 1 tsp. dried basil leaves. Do not use dried basil powder.

Directions:

1.   Heat oil in a Dutch oven. Sauté garlic and onions until onions are limp, and the garlic is aromatic.
2.   Add the rest of the vegetables: bell pepper, celery, and tomatoes. Stir fry for 1 minute. Stir in the rest of the ingredients except the cashew nuts, raisins, and the seasonings (salt and pepper.) Set the flame on high heat and wait for the cooking liquid to boil. Afterwards, reduce the heat to the lowest setting. Cook the stew for another 15 minutes, covered.
3.   Remove the lid and stir in cashews and raisins. Add salt and pepper to taste. If you want a spicier stew, add more chili powder and/or Tabasco sauce. Cook for 20 minutes or until cashews are tender. Remove from flame.
4.   Just before serving, fish out the bay leaf. Serve immediately.

Minted Chickpea Curry

Ingredients:

For the stew:

2 cups	Dried Chickpeas, washed and drained well
6 cups	Water
1 tsp.	Dried Chili Powder

1 tsp.	Turmeric Powder
1 piece	Russet Potato, washed, peeled and diced (You can substitute sweet potato.)
2 pieces	Tomatoes, washed and coarsely chopped
½ cup	Coconut Milk
-	Salt to Taste

For the garnish:

¼ cup	Fresh Cilantro Leaves, washed and coarsely chopped
¼ cup	Fresh Mint Leaves, washed and coarsely chopped
1 piece	Fresh Jalapeño pepper, washed, stem and seeds removed, minced

You would also need: Slow Cooker

Directions:

1. Set the slow cooker on its lowest setting. Place the chickpeas in and pour in the water. Cook covered for 8 hours or until chickpeas are fork tender.
2. Remove the lid and place in the rest of the stew ingredients. Stir gently. Season to taste. Place the cooker's setting to medium. Let the stew cook for another hour.
3. Serve while warm. Ladle stew into soup bowls and garnish liberally with the chopped cilantro and mint leaves. Add the chopped jalapeño pepper for a spicier stew.

Ratatouille

Ingredients:

2 tbsp.	Extra Virgin Olive Oil
2 pieces	Garlic Cloves, peeled and minced
1 piece	Onion, medium-sized, peeled, sliced into wedges

1 piece	Eggplant, medium-sized, washed, stem removed, diced
1 piece	Green Bell Pepper, small-sized, washed, stem and seeds removed, diced
10 pieces	Snap Beans, washed, dried, ends and strings removed, diced
2 pieces	Tomatoes, large-sized, cut into half to remove seeds, diced
1 piece	Zucchini, medium-sized, washed, dried, ends removed, diced
1 tbsp.	Fresh Oregano Leaves, washed, stems removed, minced
1 tbsp.	Fresh Basil Leaves, washed, stems removed, minced
2 tbsp.	Red Wine Vinegar
1 can (15oz.)	Tomato Sauce
1 cup	Vegetable Stock
-	Salt to Taste
-	Ground Black Pepper to Taste

Directions:

1. In a large Dutch oven, heat the oil over medium flame. Add the onions and stir fry until onion wedges are transparent.
2. Add the garlic and cook for another minute or until the onions are limp. Add the rest of the vegetables, except the snap beans. Pour in the red wine vinegar and stir once. Place the lid on the Dutch oven and allow the vegetables to cook for 10 minutes over low flame.
3. After 10 minutes, toss in the snap beans, tomato sauce and vegetable stock. Return heat to medium flame. Cook for another 2 minutes, uncovered.
4. Season vegetable stew with basil, oregano, salt and pepper. Adjust to taste. Remove Dutch oven from flame. Serve stew immediately.

Simple Black Bean Stew

Ingredients:

2 cups	Dried Black Beans, washed and drained well
6 cups	Water
½ cup	Light Soy Sauce
2 tbsp.	Toasted Sesame Oil
-	Salt to Taste
-	Pepper to Taste

You would also need: Slow Cooker

Directions:

Except for the salt and pepper, place everything in the slow cooker. Set to the lowest setting and cook covered for the next 8 hours. Just before serving, season with salt and pepper, if needed.

Spicy Green Bean Stew

Ingredients:

2 tbsp.	Extra Virgin Olive Oil
2 pieces	Onions, large-sized, peeled and coarsely chopped
1 pound	Green Beans, washed, dried, ends and strings removed, chopped into 2 inch slivers
1 can (28oz.)	Whole Tomatoes
2 pieces	Zucchini, small-sized, washed, ends removed, chopped into 2 inch long slivers
1 piece	Hannah sweet potato, large-sized, washed, peeled and cubed (You can substitute: Japanese sweet potato or jewel yam. Russet potato can also be used in a pinch.)
¼ cup	Water (or more, as needed
-	Dried Cayenne Pepper Powder to taste
-	Salt to taste
-	Black Pepper to Taste
½ cup	Flat Leaf Parsley, washed, stems removed, coarsely chopped, for garnish

Directions:

1. In a thick bottomed skillet, heat oil over medium flame. Sauté the onion until it turns transparent. Carefully stir in the contents of the canned whole tomatoes. Gently break the whole tomatoes into smaller pieces using your cooking spoon.

2. Add the sweet potato to the skillet. Pour in the water. Season to taste. Use more cayenne pepper powder if you want a spicier stew.

3. Turn the heat up to the highest setting to let the stew boil. Once it starts to boil, put a lid on the skillet and then turn the flame down to the lowest setting. Let the sweet

potatoes cook thoroughly. This should take between 35 to 45 minutes.

Note: After 15 minutes, check to see if there is still enough cooking liquid in the skillet. If the stew is becoming too dry, simply add more water ½ cup at a time. Pierce the sweet potato with a fork to see if it is done.

4.   When potatoes are cooked, add the zucchini and the green beans. Stir and allow to simmer for the next 7 to 10 minutes, covered. Check to see if the zucchini can be pierced through after 7 minutes. If so, remove the stew from the heat immediately. If not, let it cook some more. Again, keep a careful eye on the cooking liquid. Add more water, but only if the stew is too dry. Season to taste again.

5.   To serve: ladle hot stew into individual bowls. Garnish generously with chopped parsley leaves, and pair this with freshly made flat bread.

Tofu in Miso Soup

Ingredients:

6 cups	Vegetable Stock
¼ cup	Brown or Red Miso Paste*
1 knob	Ginger, thumb-sized, peeled and grated, stringy bits removed
20 pieces	Snow peas, washed, ends and strings removed
4 pieces	Fresh Shitake Mushrooms, large-sized, stems removed, caps sliced thinly**
1 piece	Carrot, medium-sized, washed, peeled, grated
2 pieces	Green Onions or Leeks, washed, roots trimmed, minced
1 piece	Red Bell Pepper, small-sized, washed, stem and seeds removed, minced

1 tsp.	Sesame Oil
Dash	Dried Red Pepper Flakes
1 pack	Extra Firm Tofu, rinsed gently under running water, drained and sliced into ½ inch cubes
(12.3oz)	
-	Cilantro, washed, stems removed, minced, for garnish
-	Salt to Taste
-	White Pepper Powder to Taste

*You can easily buy this in the Asian section of your local grocery store. Do not use the instant or powdered miso.

**You can substitute dried shitake mushrooms but you need to soak these overnight before using.

Directions:

1. In a small bowl, mix the miso paste with 1 cup of vegetable stock. Do this until you have a smooth but runny paste. Pour this, along with the rest of the vegetable stock and the grated ginger into a Dutch oven. Set this over high flame and bring to a boil.
2. Add all the vegetables and mushrooms. Reduce heat to the lowest setting. Cover Dutch oven with a lid and cook for 20 minutes, or until carrots can be pierced with a fork.
3. Stir in the red pepper flakes and the sesame oil. Season with salt and pepper.

To serve. Place a few cubes of tofu into individual bowls. Ladle steaming hot miso soup on top and garnish with chopped cilantro. Serve immediately.

CHAPTER 13- VEGAN BURGER RECIPES

Black Bean and Potato Patties

Ingredients:

4 pieces	Red or Russet Potatoes, large-sized, skin left intact, sliced into large chunks, boiled until fork tender, cooled completely at room temperature or chilled (You can do this in advance.)
1 pack (12.3oz)	Extra Firm Tofu, drained and pressed to remove excess liquid
1 can (15oz)	Black Beans, washed thoroughly under running water, drained
¼ cup	Peanut Butter
4 stalks	Scallions, washed, roots trimmed and minced
2 tbsp.	Dried Red Pepper Flakes
2 pieces	Garlic Cloves, peeled and minced
1 tsp.	Green Curry Paste
1 tsp.	Sriracha Sauce or any hot sauce you prefer
¼ tsp	Dried Coriander Powder
¼ cup	Whole Wheat Pastry Flour
-	Salt to Taste
-	Pepper to Taste

You would also need:	Oven
	Non-Stick Baking Sheet/Silicone Mat on a regular baking sheet

Fork or Masher

Aluminum Foil

Spatula

Directions:

1. Preheat your oven to 350°F or 175°C.
2. Using your fingers, break the tofu into smaller chunks into a large bowl. Add the drained beans, curry paste, garlic, peanut butter, red pepper flakes, scallions, Sriracha sauce and coriander powder. Mix well. Season with salt and pepper.
3. With a fork or masher, mash potatoes in a separate bowl. Then transfer the lot to the bowl with the black beans. Knead mixture until everything is well incorporated. Add the flour ¼ cup at a time. What you need is a mixture that is slightly wet and can easily be formed into patties. If the mixture is too dry or has too much flour, it will crumble when you bake it. You do not have to use all the flour.
4. Form 10 patties. Place these on the baking sheet and cover with aluminum foil. Bake one side for 15 minutes. Afterwards, remove the baking sheet from the oven, flip the burgers with a spatula and return to cook for another 15 minutes, covered. In the last 3 minutes of its baking time, carefully remove the foil to give the patties a brown outer layer. Let this rest for 2 minutes out of the oven before serving.

Masala Burger

Ingredients:

1 cup	Fully Cooked Lentils
2/3 cup	Water
2 tbsp.	Olive Oil

1 tsp.	Garam Masala
½ tsp.	Cumin Powder
-	Salt to Taste
-	Pepper to Taste
-	Vegetable Oil For Frying

You would also need: Skillet

Paper Towels

Directions:

1. Mix all the ingredients together, except for the frying oil. Season well with salt and pepper. Form into 4 patties.
2. Set the skillet over high flame. Wait for the oil to heat up a bit. Add enough vegetable oil to generously coat the bottom of the pan. Gently slide in the patties and fry for 5 minutes on the first side, and 3 minutes on the other. Transfer cooked patties onto paper towels to remove excess oil. Serve immediately.

Three Pepper Burger Patties

Ingredients:

2 tbsp.	Vegetable Oil
6 pieces	Shallots, medium-sized, peeled and minced
6 pieces	Canned Portobello or Button Mushroom, large-sized, washed under running water, drained, minced
1 piece	Green Bell Pepper, small-sized, stem and seeds removed, minced
1 piece	Red Bell Pepper, small-sized, stem and seeds

removed, minced

1 piece	Yellow or Orange Bell Pepper, small-sized, stem and seeds removed, minced
3 cups	Cooked Brown Rice, cooled completely
2 tbsp.	Peanut Butter
3 tbsp.	Sesame Oil
1 tsp.	Cumin Powder
1 tbsp.	Sriracha Sauce or any hot sauce you prefer
-	Salt to Taste
-	Cornstarch (optional)

You would also need: Wok

Oven

Baking Sheet with nonstick surface or silicone mat

Saran Wrap

Spatula

Directions:

1. Over high flame, preheat the wok until it is slightly smoky. Stir fry the 3 bell peppers, mushrooms and onions for 3 minutes. Turn heat off.
2. Toss the cooked rice, cumin powder, peanut butter, sesame oil and the Sriracha sauce into wok. Make sure that everything is well distributed. Allow this to cool completely at room temperature.
3. Transfer the cooled pepper-mushroom mix to a glass bowl and cover with a saran wrap. Refrigerate this for at least 20 minutes before shaping the mixture into patties. It would be preferably though if this could be chilled overnight.

4. When you are ready to bake, preheat oven to 350°F or 175°C. Spray cooking oil on a non-stick baking sheet, or on the silicone baking mat. (You can also use parchment paper on a cookie sheet.)
5. Form the pepper-mushroom mixture into patties. If the patties are too wet, add a tablespoon of cornstarch at a time. Mix well. This batch should yield 8 burgers.
6. Place burger patties onto the baking sheet and slide into the oven to cook for 15 to 20 minutes, uncovered. Using a spatula, flip the burgers and bake for another 15 minutes. Once cooked, remove baking sheet from the oven, and allow to cool slightly before serving.

Wheat Free Sweet Potato Burger

Ingredients:

6 pieces	Sweet Potatoes, large-sized, washed, peeled and julienned, soaked in cold water to prevent discoloration, drained well before using, squeeze out excess moisture using a tea towel
1 piece	Onion, large-sized, peeled and minced
1 piece	Green or Red Bell Pepper, medium-sized, washed, stem and seeds removed, minced
1 piece	Yellow or Orange Pimiento, small-sized, washed, stem and seeds removed, minced
3 cups	Chickpea Flour (You can also substitute cornmeal or oat flour.)
1 cup	Cashew Sour Cream or any non-dairy sour cream substitute (You can substitute ordinary sour cream in a pinch.)
-	Salt to Taste

- Pepper to Taste

- Vegetable Oil For Frying

You would also need: Tea Towel

Paper Towels

Directions:

1. In a large mixing bowl, combine bell pepper, onions, pimiento, sour cream, with the drained sweet potatoes. Season well with salt and pepper.
2. Incorporate the chickpea flour one cup at a time. Mix well with each addition. This should yield a very wet mixture. Adjust the taste as you go.
3. Meanwhile, heat a skillet with enough cooking oil to generously coat the bottom of the cooking surface. On medium heat, allow the oil to become slightly smoky.
4. With floured hands, scoop a handful of the sweet potato mix and form into patties. Drop patties in the heated oil and fry for 3 to 5 minutes. Cooking time would depend on how thick your patties are.
5. Place cooked patties on absorbent paper towels to remove excess oil. You can serve this with your favorite vegetable salad and/or bread, with slices of fresh tomatoes and cucumber on the side.

CHAPTER 14- VEGAN-FRIENDLY DESSERTS

Apple Cake

Ingredients:

For the cake:

1½ cups	All-Purpose Flour, Sifted
2 tsp.	Baking Powder
½ tsp.	Baking Soda
1 cup	Granulated Sugar (You can substitute any baking-friendly artificial sweetener of your choice.)
½ cup	Olive Oil
½ cup	Freshly Squeezed Apple Juice, add the pulp if possible
2 tbsp.	Maple Syrup
2 tsp.	Apple Cider Vinegar
1 tsp.	Ground Cinnamon Powder
1 tsp.	Zest of Freshly Grated Orange
5 pieces	Apples, medium-sized, washed, cored, diced
½ tsp.	Ground Nutmeg Powder
½ cup	Raw Walnuts, coarsely chopped (optional)
¼ cup	Dark Raisins (optional)

- Oil for brushing

- Flour for dusting

You would also need: Oven

9 inch Spring Form Pan

Wire Whisk

Wooden Spoon

Wire Rack

Directions:

1. Preheat the oven to 350°F or 175°C.
2. Lightly oil the sides and bottom of the spring form pan. Lightly dust the oiled surfaces with flour. Set aside.
3. In a large mixing bowl, sift the baking powder, baking soda and the flour together.
4. In another bowl, combine the sugar and the olive oil and whisk until you have a smooth consistency. Whisk in the apple juice, cinnamon powder, maple syrup, vinegar and the orange zest. Add this to the flour mixture using a wooden spoon. Mix just enough to incorporate the wet ingredients with the dry.
5. Fold in the apples and the nutmeg powder. Fold in also the raisins and walnuts if you are using these.
6. Pour the cake batter into the prepared spring form pan. Tap the pans lightly on the kitchen counter to let out the air bubbles. Bake for an hour or until an inserted toothpick at the center of the cake comes out clean.
7. Remove cake from oven, and place spring form pan on wire rack. After 30 minutes, remove the spring form pan, and set the cake on the wire rack to help it cool further. If possible, chill the cake before slicing and serving.

Banana Berry with Cashew Crisps

Ingredients:

2 cups	All-Purpose Flour
1 tsp.	Baking Powder
½ tsp.	Baking Soda
½ tsp.	Ground Cinnamon Powder
½ tsp.	Kosher or Sea Salt
4 pieces	Overripe Bananas, large-sized, peeled, flesh mashed with a fork
¾ cup	Brown Sugar (You can substitute any baking-friendly artificial sweetener of your choice.)
1/3 cup	Granulated Sugar (You can substitute any baking-friendly artificial sweetener of your choice.)
½ cup	Olive Oil
¼ cup	Unsalted Cashew Nut Butter (You can substitute almond or macadamia nut butter.)
1 tbsp.	Soy Milk
1 tsp.	Almond Extract
¾ cup	Raw Cashews, coarsely chopped
1/3 cup	Dried Cranberries
-	Oil for Brushing

You would also need: Oven

                        3 Cookie Sheets or Silicone Mat

Wooden Spoon

Wire Rack

Wire Whisk

Directions:

1. Preheat the oven to 300°F or 175°C.
2. Lightly brush 3 cookie sheets with oil. Set aside.
3. In a large mixing bowl, combine baking powder, baking soda, cinnamon powder, flour and salt. Mix well with a wooden spoon. Make a well in the center of the flour mixture.
4. In another bowl, whisk together the mashed bananas, the 2 types of sugar, olive oil, and nut butter. Try to incorporate as much air as possible to the mixture to give the cookies a lighter consistency. Whisk in the almond extract and the soy milk.
5. Pour the liquid ingredients into the center of the flour mixture. Use the wooden spoon to mix everything. Fold in the cashews and the cranberries.
6. Scoop a tablespoon of the dough and drop this on the oiled baking sheets. Leave enough spaces in between, about 3 inches apart. This would allow the cookies to spread out a little during the baking process.
7. Bake in the hot oven for 10 to 15 minutes, or until the edges of the cookies turn golden brown. Remove baking sheets from the oven. Cool slightly. Using a spatula, scoop each cookie and place it on the wire rack to cool. Serve warm.

Carrot Spice Cake

Ingredients:

1-1/3 cups	All-Purpose Flour
1 tsp.	Baking Powder

½ tsp.	Baking Soda
½ tsp.	Rock Salt or Kosher Salt
1 cup	Granulated Sugar (You can substitute any baking-friendly artificial sweetener of your choice.)
½ cup	Olive Oil
1/3 cup	Soy Milk (You can substitute your choice of dairy-free milk)
¼ cup	Apple Sauce
2 tsp.	Ground Cinnamon Powder
1 tsp.	Vanilla Extract
½ tsp.	Ground Nutmeg Powder
2 pieces	Carrots, large-sized, washed, peeled and finely grated
1 cup	Raw Walnuts, coarsely chopped (Optional)
-	Oil for Brushing
-	Flour for Dusting

You would also need:	Oven
	9 inch Spring Form Pan
	Wire Whisk
	Wooden Spoon
	Wire Rack

Directions:

1.   Preheat the oven to 350°F or 175°C.

2. Lightly oil the sides and bottom of the spring form pan. Lightly dust the oiled surfaces with flour. Set aside.
3. In a large mixing bowl, sift the baking powder, baking soda and the flour together. Add the salt and mix with a wooden spoon.
4. In another bowl, combine the sugar and ½ cup of olive oil. Beat with a wire whisk until foamy. Pour in the applesauce, cinnamon powder, nutmeg, soy milk and vanilla extract. Mix well.
5. Make a well in the center of the flour mixture. Pour the liquid ingredients into the well. Mix with a wooden spoon until just combined. Do not over mix. Gently fold in the carrots and the walnuts if you are using these.
6. Pour cake batter into the prepared spring form pan and bake for 50 minutes, or until toothpick inserted at the center of the cake comes out clean. Remove spring form pan from the oven.

    Place this on a wire rack to cool completely at room temperature, about an hour. Remove cake from spring form pan and the place on the wire rack for another 15 minutes. Serve immediately.

Corn and Berry Muffins

Ingredients:

1 ¼ cup	All-Purpose Flour
¾ cup	Coarse Cornmeal
1/3 cup	Granulated Sugar (You can substitute any baking-friendly artificial sweetener of your choice.)
2 tsp.	Baking Powder
½ tsp.	Baking Soda
¼ tsp.	Rock or Sea Salt

Samantha Michaels

1 cup	Soy Milk
1/3 cup	Canola Oil
1 tbsp.	Orange Zest, Freshly Grated
1 cup	Fresh or Frozen Blueberries, washed under running water, drained well (You can use other berries of your choice.)

You would also need: Oven

12 cup Muffin Pan with paper liners

Wooden Spoon

Wire Rack

Directions:

1. Preheat the oven to 400°F or 200°C.
2. Line the muffin cups with paper liners.
3. In a large mixing bowl, combine baking powder, baking soda, cornmeal, flour, salt and sugar. Mix well, then make a well in the center of the flour mixture.
4. Pour the canola oil, orange zest and soy milk in the well. Stir with a wooden spoon. Very gently, fold in the berries.
5. Spoon cake batter into the muffin cups. Fill the muffin cups only halfway up.
6. Bake in the hot oven for 18 to 20 minutes, or until the surface of the muffins turn golden brown. Remove muffin pan from the oven. Carefully remove each muffin and place on the wire rack to cool completely at room temperature, or about 30 to 45 minutes.

Chocolate Pumpkin Muffins

Ingredients:

1 cup	All-Purpose Flour, sifted
1 tsp.	Baking Powder
½ tsp.	Baking Soda
1 tsp.	Ground Cinnamon Powder
1 can (16oz)	Pure Pumpkin
1 cup	Granulated Sugar (You can substitute any baking-friendly artificial sweetener of your choice.)
½ cup	Canola Oil
1/3 cup	Soy Milk
¾ cup	Non-Dairy Chocolate Chips
½ cup	Chopped Walnuts (Optional)

You would also need: Oven

12 cup Muffin Pan with paper liners

Wooden Spoon

Wire Rack

Directions:

1. Preheat the oven to 350°F or 175°C.
2. Line the muffin cups with paper liners.
3. In a large mixing bowl, sift together flour, baking powder, baking soda, and cinnamon powder. Make a well at the center.
4. Add in the oil, pumpkin, soy milk and sugar. Mix well with a wooden spoon. Gently fold in the chocolate chips and the walnuts if you are using these.

5.   Spoon cake batter into the muffin cups. Fill muffin cups
        only halfway.

Bake in the hot oven for 25 to 30 minutes, or until the surface of
the muffins turn golden brown. Remove muffin pan from the
oven. Carefully remove each muffin and place on the wire rack
to cool completely at room temperature, or about 30 minutes.

CHAPTER 15- JUICES & SMOOTHIES

Banana-Strawberry Smoothie

Ingredients:

1 cup	Water (Add more if desired)
½ cup	Crushed Ice (Add more if desired)
2 pieces	Bananas, large-sized, peeled, flesh sliced in half
1 pound	Strawberries, washed, stems removed (You can substitute with other berries.)

You would also need: Blender

Directions:

Blend everything until smooth. If the drink is too thick, add ¼ cup of water and blend once more. If it is too runny, add more ice. Serve immediately.

If you prefer a sweeter smoothie, simply add one or more bananas to the recipe.

Berry Elixir

Ingredients:

1 cup	Water, add more if desired
½ cup	Crushed ice, add more if desired
1 cup	Fresh or Frozen Strawberries, washed, stems removed
1 cup	Fresh or Frozen Blueberries, washed, stems removed
1 cup	Fresh or Frozen Raspberries, washed, stems removed
1 piece	Cucumber, washed, peeled, halved, seeds scooped out
1 piece	Lemon, take as much of the zest as possible, seeds removed, squeeze as much juice and lemon as you can.

You would also need: Blender

Directions:

Blend everything until smooth. If the drink is too thick, add ¼ cup of water and blend once more. If it is too runny, add more ice. Serve immediately.

Berry Red Drink

Ingredients:

1 cup	Crushed Ice, add more if desired
½ cup	Raspberries, washed, stems trimmed, halved

½ cup	Strawberries, washed, stems trimmed, halved
½ cup	Blackberries, washed, stems trimmed, halved
1 piece	Apple, large-sized, washed, cored and cubed
1 piece	Orange, peeled, membrane and seeds removed
½ piece	Red Beet Root, washed, peeled and chopped
1 tbsp.	Freshly Squeezed Lemon or Lime Juice

You would also need: Blender

Directions:

Blend everything until smooth. Serve immediately.

Berry Infusion

Ingredients:

½ cup	Fresh Blueberries, stems removed, washed, drained well
½ cup	Fresh Strawberries, stems removed, washed, drained well
½ cup	Frozen Blackberries, stems removed, washed, drained well
½ cup	Frozen Cranberries, stems removed, washed, drained well
1 cup	Water
-	Crushed Ice
-	Mint Leaves, washed, stems removed, for garnish

Samantha Michaels

You would also need: Pitcher with Lid or Mason Jar,
thoroughly cleaned

Wooden Spoon

Directions:

1. Pour all the berries into the pitcher. Bruise these using a wooden spoon to help them release their flavor into the water.
2. Pour the water in and mix well with the wooden spoon.
3. Add enough crushed ice to fill the pitcher. Put the lid on, and let the berries' flavors infuse into the water for at least 3 hours.
4. Serve with mint leaf garnish.

Blackberry Sage Water Infusion

Ingredients:

½ cup   Fresh sage leaves, washed, squeeze dried

2 cups   Fresh blackberries, washed, halved

2 cups   Water

-            Crushed Ice

Sprig   Sage Leaf for garnish

You would also need: Pitcher with lid or mason jar,
thoroughly cleaned

Wooden Spoon

Directions:

1. Place the raspberries into the pitcher. Mush these slightly using the wooden spoon.
2. Place the sage into the pitcher. Bruise these using the wooden spoon.
3. Add the water into the pitcher and stir well.

4. Add enough crushed ice to fill the pitcher. Put the lid on, and let the flavors infuse into the water for at least 3 hours.
5. Serve with sage leaf garnish.

Citrus Water

Ingredients:

½ piece	Grapefruit, washed, quartered, visible seeds removed
1 piece	Lemon, washed, quartered, visible seeds removed
1 piece	Lime, washed quartered, visible seeds removed
1 piece	Orange, washed, quartered, visible seeds removed
1 cup	Water
-	Crushed Ice
-	Mint Leaves, washed, stems removed, for garnish

| You would also need: | Pitcher with lid or mason jar, thoroughly cleaned |
| | Wooden Spoon |

Directions:

1. Squeeze the juices of the citrus fruits into the pitcher, then throw in the quartered slices. Try to fish out as much seeds as possible.
2. Pour the water in and mix well with the wooden spoon.
3. Add enough crushed ice to fill the pitcher. Put the lid on, and let the fruits' flavors infuse into the water for at least 3 hours.

4.   Serve with mint leaf garnish.

Cranberry Slush

Ingredients:

1 cup	Water, add more if desired
½ cup	Crushed Ice, add more if desired
3 pieces	Bananas, large-sized, peeled, flesh sliced in half
1 cup	Fresh Cranberries, washed, stems removed

You would also need: Blender

Directions:

Blend everything until smooth. If the drink is too thick, add ¼ cup of water and blend once more. If it is too runny, add more ice. Serve immediately.

If the slush is too tart, either add more bananas to the recipe, or lessen the amount of cranberries.

Fruity Coconut Mix

Ingredients:

½ cup	Crushed Ice, add more if desired
2 cups	Seedless Grapes, washed, stems removed (You can use any variety of grapes as long as you carefully remove the seeds within.)
1 piece	Apple, large-sized, washed, cored and coarsely chopped
1 piece	Orange, medium-sized, washed, peeled,

membrane and seeds removed, include juice

1 piece   Young Coconut, scrape the meat and juice

You would also need: Blender

Directions:

Blend everything until smooth. Add more ice if the drink is too runny. Serve immediately.

Grapefruit Elixir

Ingredients:

1 cup   Water, add more if desired

½ cup   Crushed Ice, add more if desired

1 cup   Fresh Grapefruit Juice, seeds removed, squeeze as much juice and pulp as you can

1 cup   Fresh or Frozen Berries of your choice, stems removed

You would also need: Blender

Directions:

Blend everything until smooth. If the drink is too thick, add ¼ cup of water and blend once more. If it is too runny, add more ice. Serve immediately.

Mango Kale Smoothie

Ingredients:

1 cup   Water, add more if desired

½ cup	Crushed Ice, add more if desired
4 pieces	Fresh Kale Leaves, large-sized, washed, dried and torn
2 pieces	Flesh Scooped Out

You would also need: Blender or Juicer

Directions:

Blend everything until smooth. If the drink is too thick, add ¼ cup of water and blend once more. If it is too runny, add more ice. Serve immediately.

Minty Pineapple Water

Ingredients:

2 cups	Fresh Pineapple Chucks, cored, reserve as much as of its juice as possible
1 Sprig	Mint, washed, root and other woody bits trimmed off
1 cup	Water
-	Crushed Ice

You would also need: Pitcher with lid or mason jar, thoroughly cleaned

Wooden Spoon

Directions:

1. Place the pineapple chunks into the pitcher. Using a wooden spoon, try to break the chunks into smaller pieces without completely turning these into pulp.
2. Add the mint sprig and bruise some of the leaves using the wooden spoon.
3. Pour the water in and mix well.

4.   Add enough crushed ice to fill the pitcher. Put the lid on, and let the flavors infuse into the water for at least 3 hours.

Pear and Dates Drink

Ingredients:

1 cup	Water, add more if desired
½ cup	Crushed Ice, add more if desired
2 pieces	Pears, large-sized, washed, cored and cubed
2 pieces	Soft dates, pitted
¼ head	Fennel, washed, roots trimmed, coarsely chopped
¼ tsp.	Vanilla extract

You would also need: Blender

Directions:

Blend everything until smooth. If the drink is too thick, add ¼ cup of water and blend once more. If it is too runny, add more icc. Serve immediately.

Raspberry Lime Water Infusion

Ingredients:

2 pieces	Fresh Limes, washed, quartered
1 cup	Fresh Raspberries, washed, stems removed, halved
2 cups	Water

-            Crushed Ice

Sprig   Mint Leaf for garnish

You would also need: Pitcher with lid or mason jar,
                                       thoroughly cleaned

                                       Wooden Spoon

Directions:

1.   Place the halved raspberries into the pitcher. Mush these slightly using the wooden spoon.
2.   Squeeze the limes into the pitcher, then throw in the spent lime quarters into the pitcher as well.
3.   Add the water into the pitcher and stir well.
4.   Add enough crushed ice to fill the pitcher. Put the lid on, and let the flavors infuse into the water for at least 3 hours.
5.   Serve with mint leaf garnish.

Spicy Blueberry and Banana Drink

Ingredients:

1 cup   Water, add more if desired

½ cup   Crushed Ice, add more if desired

2 cups   Fresh or Frozen Blueberries, washed, stems removed

2 pieces   Bananas, large-sized, peeled, flesh sliced in half

¼ tsp.   Ground Cardamom Powder

1/8 tsp.   Ground Cinnamon Powder

You would also need: Blender

Directions:

Blend everything until smooth. If the drink is too thick, add ¼ cup of water and blend once more. If it is too runny, add more ice. Serve immediately.

Spiked Apple Juice

Ingredients:

1 cup	Water, add more if desired
1 cup	Crushed Ice, add more if desired
2 pieces	Apples, large-sized, washed, cored and cubed
1 piece	Carrot, large-sized, washed, peeled and cubed
1 piece	Celery Stalk, medium-sized, washed, coarsely chopped
2 tbsp.	Freshly squeezed lemon juice, use as much of the pulp as possible

You would also need: Blender

Directions:

Blend everything until smooth. If the drink is too thick, add ¼ cup of water and blend once more. If it is too runny, add more ice. Serve immediately.

Spiked Green Tea Smoothie

Ingredients:

1 cup	Water, add more if desired
½ cup	Crushed Ice, add more if desired
1 cup	Green Tea or Macha Powder

1 handful	Baby Spinach Leaves, washed, dried and torn
1 piece	Apple, medium-sized, washed, cored and cubed
2 pieces	Bananas, medium-sized, peeled, flesh sliced in half
Dash	Nutmeg Powder

You would also need: Blender

Directions:

Blend everything until smooth. If the drink is too thick, add ¼ cup of water and blend once more. If it is too runny, add more ice. Serve immediately.

Vanilla Kiwi and Spinach Smoothie

Ingredients:

1 cup	Water, add more if desired
½ cup	Crushed Ice, add more if desired
1 handful	Fresh Baby Spinach, washed, dried and torn
5 pieces	Kiwi, medium-sized, washed, and peeled
2 pieces	Bananas, peeled, flesh sliced in half
½ tsp.	Vanilla Extract

You would also need: Blender

Directions:

Blend everything until smooth. If the drink is too thick, add ¼ cup of water and blend once more. If it is too runny, add more ice. Serve immediately.

Watermelon Rosemary Water Infusion

Ingredients:

¼ cup	Fresh Rosemary Leaves, washed, stems removed
½ piece	Fresh Watermelon, flesh scooped out, reserve as much of the liquid as possible
2 cups	Water
-	Crushed Ice
You would also need:	Pitcher with lid or mason jar, thoroughly cleaned
	Wooden Spoon

Directions:

1. Place the scooped watermelon and juices into the pitcher.
2. Place the rosemary leaves into the pitcher. Bruise these slightly by pounding the needles using the wooden spoon.
3. Add the water into the pitcher and stir well.
4. Add enough crushed ice to fill the pitcher. Put the lid on, and let the flavors infuse into the water for at least 3 hours.
5. Serve with sage leaf garnish.

ABOUT THE AUTHOR

Samantha Michaels has spent years helping people overcome health challenges, lose weight and reach ideal health goals while enjoying good and healthy food. She is an author of numerous health books and provide amazing yet very healthy recipes everyone can enjoy.

She loves food and spends most of her time helping people address diet challenges by teaching them to cook the right meals. Her diet programs have helped a lot of people lose weight in a smart, practical way and she lives what she preaches that you do not have to get hungry while on a diet.

Made in the USA
Lexington, KY
14 May 2017